ADVANCE ACCLAIM

"The evolution of psychotherapy points toward integration and this book leads the way."

—Bill O'Hanlon, featured Oprah author of *Do One Thing Different*

"This book provides a quantum leap into understanding how the interactions of brain, mind, and body, in an ever-changing environment, determine the meaning and quality of our existence. The inevitable results of contemplating this mind-altering synthesis is to live life more fully while positively affecting the lives of others. Everyone who cares about mental and physical health owes a huge debt of gratitude to John Arden for this brilliant contribution to the science of well-being."

—Harvey Milkman, Ph.D., Professor Emeritus, Metropolitan State University of Denver, Author of *Craving for Ecstasy and Natural Highs*, Fulbright Scholar, Reykjavik University, Iceland

"The discovery of interactions between the mind, brain, and our genes is constantly being developed. It is critical for health professionals to have a complete overview of the current research within this field. John Arden's book is an excellent source of scientific summaries and therapeutic implications. It points towards the psychotherapy of the 21st century."

—Anna Leybina, Ph.D., Director, Centre for International and Regional Projects at Civil Service and Personnel Directorate, Moscow Government

"John Arden has very creatively and in simple language represented the complex multi-directional causal relationships that explain the vast field of psychoneuroimmunology and epigenetics, along with cognitive dynamics, that provide an integrated psychotherapy approach. The simple language of the book makes it easy to comprehend and apply the information in the field of therapy. Each student of health psychology and practitioners of psychotherapy must read this book to get this holistic perspective."

—Sanjeev Sahni, Ph.D., Director of the Institute of Behavioral Sciences, O. P. Jindal Global University, India

Mind–Brain–Gene

The Norton Series on Interpersonal Neurobiology
Louis Cozolino, PhD, Series Editor
Allan N. Schore, PhD, Series Editor, 2007–2014
Daniel J. Siegel, MD, Founding Editor

The field of mental health is in a tremendously exciting period of growth and conceptual reorganization. Independent findings from a variety of scientific endeavors are converging in an interdisciplinary view of the mind and mental well-being. An interpersonal neurobiology of human development enables us to understand that the structure and function of the mind and brain are shaped by experiences, especially those involving emotional relationships.

The Norton Series on Interpersonal Neurobiology provides cutting-edge, multidisciplinary views that further our unprderstanding of the complex neurobiology of the human mind. By drawing on a wide range of traditionally independent fields of research—such as neurobiology, genetics, memory, attachment, complex systems, anthropology, and evolutionary psychology—these texts offer mental health professionals a review and synthesis of scientific findings often inaccessible to clinicians. The books advance our understanding of human experience by finding the unity of knowledge, or consilience, that emerges with the translation of findings from numerous domains of study into a common language and conceptual framework. The series integrates the best of modern science with the healing art of psychotherapy.

A NORTON PROFESSIONAL BOOK

MIND–BRAIN–GENE

Toward Psychotherapy Integration

JOHN B. ARDEN

Foreword by Louis Cozolino

W. W. NORTON & COMPANY
Independent Publishers Since 1923
New York • London

NOTE TO READERS: Standards of clinical practice and protocol change over time, and no technique or recommendation is guaranteed to be safe or effective in all circumstances. This volume is intended as a general information resource for professionals practicing in the field of psychotherapy and mental health; it is not a substitute for appropriate training, peer review, and/or clinical supervision. Neither the publisher nor the author can guarantee the complete accuracy, efficacy, or appropriateness of any particular recommendation in every respect.

Printed in the United States of America
First Edition

For information about permission to reproduce selections from this book, write to Permissions, W. W. Norton & Company, Inc., 500 Fifth Avenue, New York, NY 10110

For information about special discounts for bulk purchases, please contact W. W. Norton Special Sales at specialsales@wwnorton.com or 800-233-4830

Manufacturing by LSC Harrisonburg
Production manager: Katelyn MacKenzie

Library of Congress Cataloging-in-Publication Data

Names: Arden, John Boghosian, author.
Title: Mind-brain-gene : toward psychotherapy integration / John B. Arden.
Other titles: Norton series on interpersonal neurobiology.
Description: New York : W. W. Norton & Company, [2019] | Series: Norton series on interpersonal neurobiology | "A Norton Professional Book." | Includes bibliographical references.
Identifiers: LCCN 2018022160 | ISBN 9780393711844 (hardcover)
Subjects: | MESH: Mental Disorders—psychology | Psychotherapy—methods | Brain—physiology | Genetic Phenomena—physiology | Immune System Phenomena—physiology
Classification: LCC RC480.5 | NLM WM 420 | DDC 616.89/14—dc23 LC record available at https://lccn.loc.gov/2018022160

W. W. Norton & Company, Inc., 500 Fifth Avenue, New York, N.Y. 10110
www.wwnorton.com

W. W. Norton & Company Ltd., 15 Carlisle Street, London W1D 3BS

1 2 3 4 5 6 7 8 9 0

Dedicated to the countless healthcare professionals who have committed their lives to helping the millions of traumatized refugees.

Contents

Foreword *ix*
Acknowledgments *xiii*
Introduction *xvii*

ONE
"Self"-Organization 1

TWO
The Social Self 27

THREE
Behavior-Gene Interactions 59

FOUR
The Body-Mind and Health 91

FIVE
Self-Maintenance 115

SIX
Motivation, Habits, and Addiction 139

SEVEN
Stress and Autostress 163

EIGHT
The Trauma Spectrum 191

NINE
TRANSCENDING RIGIDITY 217

TEN
MIND IN TIME 243

References *271*
Index *295*

Foreword

As we move deeper into the 21st century, we carry with us a wealth of data, theoretical knowledge, and new technologies unimaginable just a few short years ago. As a result, we are moving past the linear logic and simplistic nature-nurture debates so prevalent during the 20th century. We now find ourselves thinking in terms of complexity, self-organizing systems, and the synergistic interactions of mind, brain, genes, and culture that create all living systems and give rise to human experience. These new discoveries are important scientific advancements that are leading us to a deeper and hopefully more accurate understanding of mental health and mental illness. It is my pleasure to welcome *Mind–Brain–Gene* into the Norton Series in Interpersonal Neurobiology as part of this important movement.

In this new book, John Arden takes a long stride forward in articulating an emerging model of clinical theory, case conceptualization, and therapeutic practice that attempts to integrate these new scientific findings into clinical practice. Multiple areas of research from genetics, epigenetics, and neuroscience, to attachment, development, and psychoneuroimmunology have demonstrated the complex interconnection taking place between our minds, brains, and genes during lifelong adaptation and change. While many are overwhelmed by the complexity of these multiple fields and retreat to narrow definitions of "how to" modes of therapy, Dr. Arden takes on these challenges with interest, enthusiasm, and compassion.

Mind–Brain–Gene is a wake-up call for those who think that knowing how to use a specific technique is enough to be an effective psychotherapist. Being a psychotherapist is far more than following a manual, assigning homework, or applying one technique. It is grounded in careful observation, active listening and learning to ask the right questions. It is knowing and being able to creatively utilize relevant information from an array of scientific disciplines. Finally, all of these things need to be grounded in an existential engagement with our clients fueled by curiosity, caring, and compassion. You will find all of these qualities embodied and expressed in Dr. Arden's examples and case studies. You will notice that while his thinking is sophisticated, his interactions and interventions are straightforward and altogether human. Knowing how to communicate with clients is just as important as having the right answers.

This book is being published at an interesting time in our field. On the one hand, advances in neuroscience and epigenetics are making it clear that psychotherapy is a powerful intervention. We are learning that the nature and quality of relationships throughout life have the ability to trigger epigenetic processes that alter gene expression in ways that can both help and hurt us. We also know that brains are adaptational processes, hubs of energy and information that can be significantly altered within healing relationships. Psychotherapy is clearly a deeply biological intervention, as well as a social and emotional one, central to the fabric of the human life. Despite what many medical professionals believe and what insurance companies profess to support their bottom lines, psychotherapy is not a second class or subordinate form of treatment. It is relevant, powerful, and makes a difference in the lives of untold numbers of people.

On the other hand, the field of psychotherapy is moving in the direction of being "paraprofessionalized" and almost anti-intellectual. An ever-growing number of training programs march students through a list of antiquated academic courses, send them to practicums where they are undertrained, and socially promoted rather than rigorously evaluated. Our clients are people in need and psychotherapy is a powerful treatment modality that needs to be taken more seriously both by society and the field of mental health.

In *Mind–Brain–Gene*, Dr. Arden is challenging us to think more broadly, more deeply, and in a more complex manner about the factors that contribute to human health throughout life and across the generations. He is summoning us to open our eyes to the broader horizons of human experience in which we and our clients live and to bring this knowledge to bear in our thinking and work. Dr. Arden is also inviting us to consider new ways of thinking and to not be frightened by what we don't know or find difficult to understand. My hope, and most likely his, is that this expanded knowledge and way of thinking will lead us to more inclusive case conceptualizations, accurate diagnoses, and efficacious treatments.

Louis Cozolino
Los Angeles, 2018

Acknowledgments

All books come to life through the contributions of many people. This book has many. First among them is Lou Cozolino. Not only did he invite me to write it, but he guided me through every step. When the manuscript was not ready for submission he gently encouraged rigorous editing, for which he offered his talents as a superb writer and scholar. His wisdom, warmth, and patience will forever be appreciated.

The insightful readers of selected chapters who offered their expert critiques and edits include: Vincent Felliti, Jodi Trafton, Paul Golding, and Rob Decker.

My wife, Vicki, maintained patience, support, and encouragement through the nearly three-year process. This, despite making a major move, many family transitions, and retiring from our respective jobs.

The team at Norton has been the most professional, sharp, and kind that I have ever experienced in any of my prior fourteen books. Led by Deborah Malmud, with support from Kate Prince, Mariah Eppes, Sara McBride, and Nicholas Fuenzalida, made my work so much easier.

Mind–Brain–Gene

Introduction

The causes and remedies to psychological disorders have long been the subject of debate and ongoing research. Every few years a seemingly new psychotherapeutic approach surges in popularity, only to fade away later as a distant memory. Recently several lines of converging research have identified the wide variety of interactive factors underlying many health and mental health problems. This book synthesizes the already substantial literature on psychoneuroimmunology and epigenetics, combining it with the neuroscience of emotional, interpersonal, and cognitive dynamics, with psychotherapeutic approaches to offer an integrated vision of psychotherapy. The integrative model promotes a sea change in how we conceptualize mental health problems and their solutions.

We explore the multidirectional causal relationships among stress, depression, anxiety, the immune system, and gene expression. The interaction among all these factors has been illuminated by studies examining the effects of lifestyle factors on the incidence of health and psychological problems. There are significant relationships among immune system functions, stress, insecure attachment, anxiety, depression, poor nutrition, poor-quality sleep, physical inactivity, and neurophysiological dysregulation. For example, insecure attachment, deprivation, and child abuse contribute to anxiety and depression in far more extensive ways then was once believed. A complex range of health conditions affect millions of people who seek psychotherapy.

Because we have reached the point in the evolution of psychotherapy where the dividing line between mental health and physical health care has evaporated, we must address the overall health of clients. Physical health and mental health are not just two sides of the same coin as if connected but still discrete dimensions; within and between each dimension are multiple nonlinear feedback loops that constantly affect all aspects of health. An integrated approach necessitates addressing the interdependence of health and behavior in psychotherapy.

Given that psychotherapy will increasingly address health factors, this integration brings those fields of research that had previously been compartmentalized into a robust vision of psychotherapy. I apply this approach to psychological disorders such as anxiety and depression to bring the integrative model to life.

Thus, this book presents a perspective antithetical to reductionism. It explores how the immune system can become dysregulated, how autoimmune disorders are associated with mental health problems, and how those same mental health problems feed back in nonlinear causation to other health problems. For example, while type 2 diabetes increases depression, so too depression exacerbates diabetes. Nor can all mental phenomena be reduced to the neuron or a specific neural network. Thinking, emotion, and behavior feed back to produce brain change. The feedback loops within the mind-brain-gene continuum generate multidirectional causal relationships among the mind, the brain, and genetics as a larger system, not separate entities. Taken together, all these dimensions interact as nonlinear, multidimensional evolving phenomena.

Before the development of the field of psychoneuroimmunology, the mind, the brain, the body, and genes were assumed to function as semi-independent systems with only indirect influences upon one another. The fields of medicine and psychology segregated into specialties, and providers practiced health care separately from one another. They may refer their patients to a specialist in a particular area outside their own narrowly defined expertise to "fix the problem" they had little knowledge of how to address.

Psychotherapy and health care in the twenty-first century are interdependent. No longer can we justify compartmentalizing the domains to specialists by rationalizing that a particular problem is "not my specialty." Sure, there are and will continue to be health care providers whose experience and training exceed others in particular areas—I would not go to a dermatologist for a second opinion about my heart, neither would I go to a cardiologist about a bunion. My point is that psychotherapists need to know about the factors that are common to poor mental health and poor physical health.

No longer are the factors influencing mental health and general health causes thought to occur through a linear bottom-up causal chain of reactions starting with "bad genes" causing psychological disorders, nor is the semidualist concept that causality is merely a top-down process. The truth is that we are not just the result of top-down and bottom-up factors but also side to side in myriad complex interactions.

During the past few decades efforts have been made to bridge the gaps between mind and brain. The pioneering work by Allan Shore (1999), Dan Siegel (2015), and Lou Cozolino (2014) has bridged the gap between attachment research and interpersonal neurobiology. Ernie Rossi (2002) has long promoted the concept that psychotherapy contributes to gene expression. Lou Cozolino (2014) and I (Arden 2009, 2015) have integrated neuroscience with psychotherapy. In the field of psychoneuroimmunology George Solomon (1987) pioneered the focus on the interactions among the mind, brain, and immune system. This book attempts to integrate all these areas and combine them with the research on habit, meditation, self-care behaviors, and the mental operating networks.

COMPLEXITY

To understand how the multiple factors that contribute to mental health coevolve, we need an interdisciplinary perspective. Complexity theory offers this fresh and inclusive perspective. This theory grew out of systems theory and draws on the natural sciences to explain multilevel interactions and feedback loops.

Feedback loops help promote growth and maintain homeostasis. We cannot become more resilient and durable without them. Positive feedback loops promote growth and boundary expansion, critical for a person's development. Negative feedback loops help regulate stasis, like a thermostat helping regulate the temperature in the room, or stress tolerance for that person. The dynamic interplay between positive and negative feedback loops is affected by changes in the environment that we encounter and often create. Psychotherapy promotes positive and negative feedback loops to enhance mental health.

These complex feedback loops cannot be reduced to any one factor alone within the mind-brain-gene spectrum. They cause and are caused by each other. And within each subsystem there are multiple feedback loops that maintain stability and change as we adapt to changing conditions in our lives. As such, they are self-organizing systems within the collective interactions that we call a "self." In this sense, the self is not reducible to the sum its parts; it represents a "self"-organizing process (Arden, 1996). The primary need of complex adaptive systems, such as ourselves, is that we must maintain open interaction with the environment to maintain coherence, complexity—in short, health. As complex adaptive systems we are self-organizing, feeding on biopsychosocial interactions to increase our complexity and adaptability.

We are the result of multiple interacting systems that self-organize through nonlinear leaps to higher or lower levels of organization. "Self"-organization represents self-referential coherence facilitated by the emergence of the mind maintaining openness and durability for of our individuality and constantly evolving interaction with other selves.

The mind is a meaning-making process, both conscious and nonconscious. The mental operating networks construct meaning through their nonlinear interactions and the memory systems. A person's sense of herself on social, psychological, and biological levels provides the context and meaning through which she experiences the world. These interacting systems are both in flux and stable, such that her mind embraces and promotes the dynamic stability of her sense of self. Psychological disorders form when our dynamic stability breaks down. Whether clients suffer from the emotional chaos of anxiety or the emotional rigidity of depression, psychotherapy aims to help them regain a balance between flexibility and stability.

CHAPTER SUMMARY

The following chapters cover the feedback loops comprising our sense of self and its breakdown. Please note that the cases I describe in this book, although derived from actual cases in my practice, are composites—they do not represent actual sessions, and they use fictitious names. They nonetheless exemplify approaches and dialogue I have found effective in my practice.

Chapter 1: "Self"-Organization

How does the feeling of individuality emerge? How do body sensations, emotion, and cognition interact to self-organize into the emergence of self? Chapter 1 explores the emergence of subjectivity and emotion by describing the mind's operating networks: the salience, executive, and default-mode networks. The salience network, sometimes called the "feeling network" or "the material me," involves the somatic sensation and the emergence of emotion. The executive network, sometimes called the central executive or the CEO of the brain, includes working memory, being present in the moment, and complex decision making. And the default-mode network, sometimes called the "story brain," involves self-referential thought, fantasy, daydreaming, and rumination, often about other people and our place within our relationships. It draws up episodic memories and reflects on the past as well as projects into the future. Finally, the interactions and dysregulations of implicit and explicit memory are described as self-referential information systems. They form the fabric of the self and undergo updating based on experience. The mind's operating networks use information from the long-term memory systems for continuity, stability, and "self"-organization.

Chapter 2: The Social Self

Building on interpersonal neurobiology, this chapter discusses how we thrive when nurtured or develop psychological disorders when we are not. Kindling the social brain networks in psychotherapy promotes not only mental health but also general health. The mind's operating networks may develop dynamic balance and stability or become imbalanced in response to our relationships. Positive relationships are vital for health as evidenced by impaired cardiovascular reactivity, blood pressure, cortisol levels, serum cholesterol, and natural killer cells. Research shows that early deprivation undermines affect stability later in life; people who are lonely or maintain unhealthy relationships develop neurocognitive problems associated with inflammation and abnormal neurochemistry.

Close relationships support longevity as measured by the brain's capacity for neuroplasticity and neurogenesis. Cultivating secure relationships, such as through psychotherapy, promotes stress tolerance by building feedback loops in the neuroendocrine and autonomic nervous systems that minimize anxiety and lift depression. Psychotherapy can work within these social brain systems to build or rebuild the capacity for regulatory affect to deal with the challenges of daily life. By empathetically kindling the social brain networks, therapists can help clients develop security and thrive interpersonally.

Chapter 3: Behavior-Gene Interactions

This chapter begins with the exploration of the health and mental health ramifications of the Adverse Childhood Experiences (ACE) study. This and other studies of its type highlight the interaction between early adversity and epigenetic affects expressed later in life in ill health and mental health. The rapidly evolving field of epigenetics reveals how gene-environment interactions bring about the expression or suppression of the genes. While the suppression of genes can occur through the process of methylation, gene expression can occur through a process called acetylation. Of critical importance to an integrative approach to psychotherapy, factors such as attachment, adversity, and lifestyle behaviors can have a major impact on the expression or suppression of genes.

Parental or caregiver neglect and adverse childhood experiences have been shown to suppress genes regulating cortisol receptors in the hippocampus, making it more difficult to turn off the stress response of the hypothalamic-pituitary-adrenal axis later in life. With a less viable negative feedback mechanism in place, stress tolerance is diminished. In the extreme, low cortisol receptors are associated with suicide. Women who initially experienced poor attachment early in life tend to have decreased estrogen receptors later in life and are less attentive to their own babies. On the other hand, people with secure attachment have higher levels of cortisol receptors and are better able to deal with stress. This negative feedback mechanism works effectively to increase stress tolerance. Because therapists can help clients turn on and off genes through changing their behavior, anxiety and/or depression can be more effectively addressed therapeutically.

Another way that behavior affects genetic processes in response to the environment and behavior is at the telomere level. Telomeres, which comprise the ends of the linear chromosomes, generally shorten with cell division, age, and illness. The availability of an enzyme that adds nucleotides to the telomeres, called telomerase, has been linked to health and psychological factors. While stress and depression have been associated with shorter telomere length, specific lifestyle behaviors have been associated with longer telomere length.

Chapter 4: The Body-Mind and Health

This chapter begins by illustrating how the interfaces between the immune system, mind, and the brain affect mental health. For example, chronic inflammation is significantly associated with a broad range of health and mental health problems, including devastating effects on mood, cognition, and social withdrawal. Inflammation can result from adversity and the several dimensions of the interfaces between mind, brain, and body. In fact, chronic inflammation is strongly associated with anxiety, depression, and cognitive deficits, including dementia. One of the routes by which inflammation occurs is by increases in pro-inflammatory cytokines. These cytokines contribute to and have a bidirectional causal relationship with anxiety, depression, and cognitive impairment by altering the levels of neurotransmitters such as dopamine, norepinephrine, and serotonin, as well as a variety of other effects.

Unhealthy lifestyle behaviors cause epigenetic effects that dysregulate the immune system, causing autoimmune disorders, depression, and anxiety. Of significant current importance to mental health providers is that the United States leads the world in obesity, autoimmune diseases, metabolic syndrome, and ranks number two in type 2 diabetes. Consistent with this pandemic is that two-thirds of the people in the United States are overweight. Numerous studies show that fat cells leach out toxic amounts of pro-inflammatory cytokines, and the associated ill health conditions are destructive to mental health. They can lead to a spectrum of symptoms, including fatigue, social withdrawal, disturbances in mood, cognition, sluggish movements, and depression, which are also strongly associated with anxiety and trauma. Therefore, psychotherapists must assess and address the psychoneuroimmunological feedback loops affecting their clients' mental health.

Chapter 5: Self-Maintenance

This chapter describes how self-care practices can undermine health and mental health, leading to anxiety and depression. Failure to maintain self-care represents more than merely symptoms of psychological problems; it often is the cause of and leads to averse epigenetic effects, including shortening of the telomeres. Psychotherapy by necessity promotes a firm foundation of a balanced diet, sleep architecture, and regular aerobic exercise. The psychoneuroimmunological effects underlying these factors either support or undermine physical and mental health.

Regular aerobic exercise is a powerful antidepressant and anxiolytic. Exercise also delays cognitive decline and dementia through a variety of processes, including significantly lower inflammation. Exercise promotes a healthy brain in many other ways, too, including the release of neurotrophic factors that promote healthy capillaries, glucose utilization, and neurogenesis.

Mental health is profoundly affected by diet. For example, regular consumption of simple carbohydrates, trans-fatty acids, and the wrong fats creates insulin insensitivity, chronic inflammation, and diminished neurotransmitter levels. A diet high in simple carbohydrates increases advanced glycation end-products, accelerating the formation of plaques and tangles in the brain. Prior to developing dementia, people become depressed and have cognitive deficits, and often seek psychotherapy. A poor diet also changes gut bacteria, which is associated with leaky gut, inflammation, and depression. Though there are literally hundreds of different types of bacteria in the gut, 90 percent fall in two broad categories: Firmicutes and Bacteroidetes. If the F/B ratio is skewed in the direction of the Firmicutes (fed by simple carbohydrates), leaky gut tends to occur, with increases in inflammation.

While sleep accounts for roughly one-third of our lives, poor-quality sleep can either destabilize mood and cognition. Poor-quality sleep dysregulates levels of stress hormones such as cortisol, which at high levels impairs the hippocampus and the frontal lobes. Adverse epigenetic effects, marked increases in inflammation, and shortened telomeres are associated with impaired sleep.

Chapter 6: Motivation, Habits, and Addiction

Adaptive and maladaptive habits are learned behaviors coded into our implicit memory systems. They affect our motivation, often involving formation of procedural and emotional implicit memory. Procedural memory is facilitated by the striatum and nucleus accumbens, so they become automatically associated with pleasure and/or the relief of discomfort.

Habits that become addictions can play a bidirectional causal relationship with anxiety and depression. From an integrative approach to addressing addictions, anxiety, and depression, it is important that we do not conceptually separate them as "diseases." The rigid categories dictated by the *Diagnostic and Statistical Manuals*, psychotherapy books, and seminars generally stay clear of addressing chemical dependency, while addictions providers defer to mental health providers for people with "psychiatric" disorders. Integrated psychotherapy goes beyond the one-dimensional conceptual frames of "dual diagnosis" and "co-occurring disorders." Addictions, also including those that are not chemical in nature, such as to gambling and computer games, hijack the dopamine circuits and the nucleus accumbens and striatum neural networks.

By understanding the neuroscience underlying habits, therapists can more effectively help clients boost motivation and overcome maladaptive habits. For example, most addictions downregulate dopamine receptors, making the range of potential positive experiences narrow to the addictive behavior. People who had experienced multiple adverse childhood events tend to have a reduced range of potentially positive experiences, which represents a setup to develop addictive behaviors. When stressed, they may engage in their go-to source of pleasure, their addiction. Expanding the range of positive behaviors expands the number of medium spiny neurons in the nucleus accumbens, making this part of the brain better able to put the brakes on an automatic habit generated by the striatum.

Chapter 7: Stress and Autostress

This chapter explains how the multidimensional stress systems have been reconceptualized. The terms *allostasis* and *allostatic load* help us more fully appreciate how resiliency and adaptability (allostasis) can be developed and the breakdown and dysfunction (allostatic load) can be minimized. Allostatic load involves the breakdown of regulatory feedback systems between health conditions and mental health, contributing to the development of anxiety and depression. The dysregulation of the sympathetic and parasympathetic systems can throw the immune system out of balance, contributing to affective and cognitive deficits. Meanwhile, dysregulation in the neuroendocrine system can further undermine the stabilizing feedback loops, resulting in first hypercortisolism then hypocortisolism with increases in inflammation, contributing to cardiovascular system damage as well as more systemic deficits to the central nervous system.

The formation of the anxiety disorders such as generalized anxiety, phobias, and panic hijacks the stress reaction systems so they get turned on inappropriately. A consistent pattern of false alarms transforms to an autostress disorder, an anxiety disorder. Stress promotes anxiety when the stress system is turned on too often and signals danger when there is none. Often people who experience multiple stressors then become hypervigilant and avoidant of the symptoms of stress. From this perspective an anxiety disorder feeds on the stress response system. Like autoimmune disorders that hijack the immune system so that it turns back on the body instead of protecting it, autostress disorders transform the stress response system into something that attacks the self rather than protecting it.

This chapter also describes how therapy can temper the hyperactivity of the fast track to the amygdala, which underlies overresponses to events with unrealistic and immediate threat. Clients can be taught to put the brakes on the fast track to the amygdala by activating functions in the prefrontal cortex, which increases the slow versus the fast track. Therapy promotes activation of the left prefrontal cortex with the engagement in approach behaviors and exposure to alleviate anxiety, depression, and promote allostasis.

Chapter 8: The Trauma Spectrum

This chapter explores a range of trauma-induced responses and therapeutic approaches while it attempts to transcend the "brand names" and theoretical cul-de-sacs to find common denominators among them. From so-called simple to complex trauma in etiology, and from hypervigilance to disassociation in response, therapeutic approaches must address the nature of the dysregulation of memory systems.

Given that people who experience a life-threatening trauma may go on to later develop posttraumatic stress disorder, it is important to explore the developmental and sociocultural factors and adverse childhood experiences that contribute to the lack of resiliency and vulnerability. A variety of epigenetic, psychoneuroimmunological factors are associated with risk. If an individual has poor social support, or on another level, if he has a smaller than normal hippocampus, maintains high levels of cortisol and norepinephrine in the evening, and has high levels of pro-inflammatory cytokines, he tends to be more vulnerable to developing posttraumatic stress disorder.

During the last few decades a variety of somatic-based therapies have emerged, such as Eye Movement Desensitization and Reprocessing, Emotional Freedom Techniques, Somatic Experiencing, and Sensorimotor Integration. They have competed theoretically with evidence-based practices, such as prolonged exposure and cognitive processing therapy. An integrative psychotherapeutic approach finds the common factors and incorporates a synthesis of exposure and somatic approaches.

Chapter 9: Transcending Rigidity

Depression is not a singular disorder with one etiology. The links between depression, inflammation, adverse childhood experiences, and early-life deprivation are interrelated with the incidence of illnesses such as diabetes and cardiovascular disease. Depressed clients with a history of early-life trauma demonstrate significantly higher levels of pro-inflammatory cytokines, such as interleukin-6, and higher activation of tumor necrosis factor.

Typically, hyperactivation of the amygdala is associated with anxiety disorders. Yet an enlarged and hyperactive amygdala is sometimes associated with depression. The activation of the amygdala appears to normalize after successful treatment for depression. Similarily, the role of corticotropin-releasing hormone in agitated depression has gained considerable attention because it is often elevated in depressed clients, as well as in suicide victims.

Complex adaptive systems are by nature open systems. We need interaction with the environment to grow and change. Closed systems, by contrast, are isolated, with no exchange of information with the environment. They are forced to feed on themselves. From a psychological perspective depression, with its associated behaviors of withdrawal, isolation, and lack of effort, may be thought of as promoting a closed system. People suffering from depression can be thought of as stuck in a kind of psychological rigidity. When there is a failure to achieve sufficient input, information, or energy, the depressed person begins to lose complexity. Through reversing these dysregulations she can develop open and activated interactions with her environment, and transcend the rigidity of depression.

Chapter 10: Mind in Time

The placebo effect represents one of the most provocative phenomena in health care, including psychotherapy. Several studies have revealed the placebo effect occuring in the brain. When patients believe that they are receiving a pain medication, endogenous opiates are released in the brain. The placebo effect illustrates how belief changes the brain and thus how we experience our body. Not only does the placebo effect represent a so-called top-down process, but also interactions with bottom-up sensations form a top-down feedback to change those sensations. For example, when we note "side effects" of a medication, we may assume that the medication is beginning to work to produce the "main effect."

The positive psychology research on forgiveness, compassion, and gratitude are explored in Chapter 10, with respective to their effect on mental health. Similarly, optimism and an attitude of acceptance are associated with resiliency. Together, these attitudinal perspectives play a major role in promoting mental health.

During the past few decades, mindfulness has been subsumed into the mainstream as well as within the "third wave" of therapies, such as acceptance and commitment therapy, dialectical behavior therapy, and mindfulness-based stress reduction. While this addition to preexistent therapies has made important contributions, there remains considerable misinformation regarding the research. Many well-meaning therapists assume that simply promoting mindfulness is the end-all. Meanwhile, millions of other potential readers turn away from mindfulness and books about it, worrying that an interest in Buddhist practices conflicts with their faith in theologies such as Christianity or Islam. There are similar methodologies within those traditions, which this chapter explores.

Meditative/mindfulness practices produce a range of profound health effects, as illustrated in a number of studies in neuroscience. To understand mindfulness, contemplative prayer, and related practices, it is important to note that for most people working memory lasts for 20–30 seconds. On the other hand, we all spend 30 percent of our waking hours in our default-mode network, daydreaming, planning for the future, or ruminating about the past. It is not that the default-mode network represents a dysfunctional process. Rather, it can be the source of creativity and healthy self-reference. This chapter describes the balance between executive, salience, the default-mode networks and how to stay in time to improve psychotherapeutic success.

The plethora of psychotherapy theories of the twentieth century, claiming the exclusive explanation and methodology, has given way to an integrated model of the twenty-first century. The following chapters will explore the many interconnected feedback loops that contribute to mental health or ill health. Research in neuroscience, epigenetics, and psychoneuroimmunology has shown that genes can turn on or off and that disease processes and mental health are significantly related to our lifestyle and experience. Our brain, immune, and stress response systems adapt to our experiences. Taken together, psychotherapy seeks to enhance mental as well as physical health. This book explores those interconnections and attempts to contribute to the integrative model of the future.

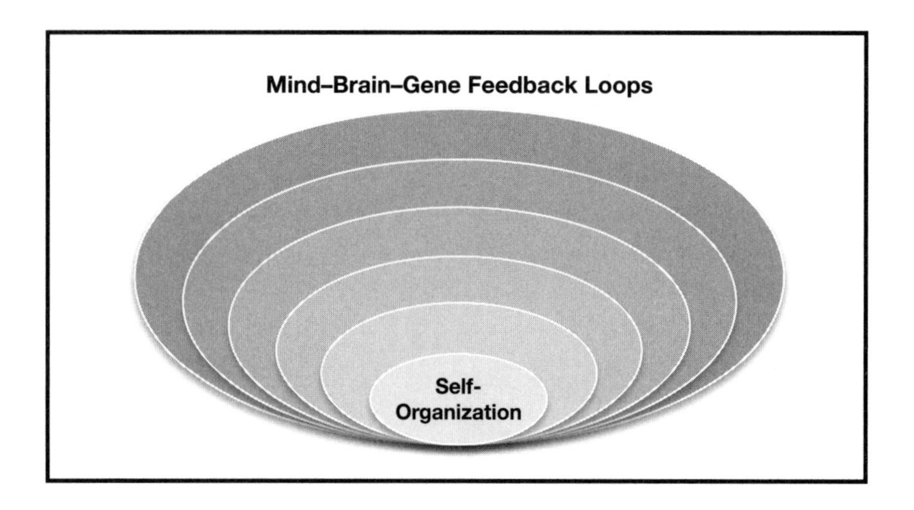
Mind–Brain–Gene Feedback Loops
Self-
Organization

ONE

"Self"-Organization

> The baby, assailed by eyes, ears, nose, skin, and entrails at once, feels it all as one great blooming, buzzing confusion.
>
> —William James

Psychotherapy in the twenty-first century is transforming its scope. Medicine, too, is becoming far more integrated. The compartmentalization of the past is dissipating as we address the complexity of the self and the feedback loops that contribute to the emergence of the mind. There is no single reason that someone becomes depressed or anxious. All of the factors that contribute to poor mental and physical health are part of a complex web of interactions and must be understood and addressed together to provide truly integrated psychotherapy.

In the movie *Analyze This*, Billy Crystal plays a psychotherapist named Doctor Sobel. In one scene a woman, named Caroline, goes on and on about a breakup with a man by saying, "He complained that I was pushing him away! But I was only trying to get in touch with my uniqueness, myself as a separate entity, as a person with my own identity. And he could not understand that!" Doctor Sobel sighs and then responds, "Caroline, things end. It is how we deal with things ending that is important." She apparently did not hear the wisdom of his feedback and goes on, "I don't understand why it's over between me and Steve. Do you think there is still hope?" To that, Doctor Sobel sighs and responds, "Well, he took out a restraining order against you, and usually that's not a good sign."

Beyond the silliness in the movie, many people seek therapy for self-discovery, to find their "true essence" or "inner self." Caroline was reaching for the archetype of the soul, the avatar, the atman, which is assumed to be the pristine "me" in each one of us. What makes you feel like "you" and me feel like "me" with unique characteristics, coherent boundaries, and separate consciousness? The idea that there is an unchanging "true self," while appealing, is an archaic concept. The self is not a static entity but a complex adaptive system that emerges from but is not reducible to its subordinate subsystems (Arden, 1996).

In the epigraph to this chapter, William James addressed how a baby is first bombarded by confusing sensations and feelings with no central sense of organization. All the sensations, including feelings generated from the immune system, gut, gene expression, and self-care behaviors, contribute to the subjective organization of the self.

This chapter explores how evolving experiences of individuality emerge from the somatic sense of self. In other words, how you feel in your body gives birth to the visceral experience of being an adaptive individual self. From there emerges what we have come to call the mind, composed of thoughts and emotions. An integrated psychotherapy must take into account how mental networks organize or fail to organize all the systems that contribute to the sense of self.

Mental operating networks that include the salience, executive, and default-mode networks serve to organize, stabilize, and provide meaning for a coherent sense of self. They are "self"-organizing networks that derive self-referential information from the implicit and explicit memory systems. Here I focus on how both dysregulation and imbalance among these networks can contribute to psychological disorders. One critical challenge of psychotherapy is to facilitate the integration of these mental networks to help the client gain a coherent and adaptive self. Emotion represents an aspect of the mind that for many people feels out control. For this reason I begin this chapter by exploring how our understanding of emotions has evolved.

BODY AND EMOTION

The paradigm shift that merged body, emotion, and cognition began with Darwin's observation that there are a common set of emotions across animal species. He described how unique similarities in emotion can be revealed by blushing in one family and that family members born blind display common facial expressions. This evolutionary perspective suggested that emotions are heritable and expressed by the body. But if human emotions are continuous with other species, do they play out in a very limited range of expression?

Some have envisioned emotions as arising from the so-called limbic system. Consistent with this belief, the antiqued triune brain theory envisioned the reptile and mammalian brains encapsulated within the cortex like Russian dolls. Similarly, Jaak Panksepp has proposed that all mammals contain seven basic affective systems: seeking (expectancy), fear (anxiety), rage (anger), lust (sexual excitement), care (nurturance), panic/grief (sadness), and play (social joy). This is a bottom-up perspective derived from extensive animal research (Panksepp & Biven, 2012). Based on a meta-analysis of neuroimaging studies, there is some support for the concept of some basic emotions among mammals (Vytal & Hamann, 2010). But how relevant is this research to humans, who have a significantly larger and more complicated cortex? And how relevant is it to psychotherapy?

Paul Ekman built on Darwin's research to develop a method to identify facial expressions associated with specific emotions, referred to as the Facial Action Coding System (FACS). Researchers and their lab associates (including my son) slowed down the videos of subjects to identify microexpressions associated with underlying emotions. The idea is that we cannot completely mask our emotions and they leak out in subtle ways. Ekman also argued that there are subtle culture-specific "display rules" concerning who can show emotion to whom and when. Though we may experience anger or fear at different extremes, there are unique differences and nuances among people. This suggests that emotions such as anger and joy combine with thoughts, which impacts how they are expressed. Other emotions, such as guilt and envy, involve even more top-down influence. And what one person may describe or feel as anger or joy may be quite different from that of another person. This suggests that psychotherapists need to identify each client's unique subtle nuances of emotion and thought and not assume that all clients experience emotions in the same way.

EMOTION: TOP DOWN AND BOTTOM-UP INTERACTIONS

If emotions are more than simply artifacts of the so-called limbic system, how are they made? Emotions are constructed by the complex feedback loops between bottom-up and top-down networks (Barnett, 2017). The feelings of individuality emerge from the self-organizing feedback loops between cortical, subcortical, and body states and the meaning-making capacity we call the mind. William James argued that emotions are generated by our physiological and behavioral responses to the environment. From this perspective, patterns of autonomic activity and behavior precede the generation and experience of emotions. For James, a bodily feeling is the emotion, and the mind uses the body as a sounding board to make meaning of what is being experienced. This bottom-up theory is characterized by the statement, "He ran from the bear because his heart pounded and his pupils dilated."

James's student Walter Cannon rejected the dominancy of bottom-up causation and instead proposed a top-down theory. Believing that thought precedes emotions, he argued that we decide what to do about body sensations such as thirst or hunger. He noted that activation of the sympathetic nervous system is too undifferentiated to identify the many gradations of human emotions. Emotionality is an intrinsic function of the brain, and emotional awareness by necessity involves the cortex (Cannon, 1939/1989). Bear families have roamed my neighborhood at night. Though I may feel like running from the bear, I may decide to stop because I know that a bear can out run me, so running is potentially more dangerous.

Bodily (interoceptive) information without top-down modulation can occur in the range of anxiety disorders, including panic disorder. Rapid heartbeat, hyperventilation, and sweating are experienced as a precursor to catastrophic outcomes. By becoming consciously aware of the body sensations and the fear that has become associated with those sensations, we can dissociate panic thoughts from rapid heartbeat. This is the crux of the exposure paradigm. Exposure exercises form part of the foundation of therapy for people with anxiety disorders: a modified top-down response is applied as the client deliberately stirs up bottom-up sensations.

Of course, most people do not suffer from panic disorder, and top-down modulation helps us remember what is worthy of our concern. James acknowledged that bottom-up feedback can be altered by top-down-directed behavior, as described in his famous phrase, "Whistling to keep up courage . . . go through the outward motions of those contrary dispositions we prefer to cultivate" (James, 1884, p. 198). The colloquial version of this top-down shaping has been promoted by the twelve-step programs with the sayings "Fake it until you make it" and "Act as if."

Fear is not an encapsulated emotion or a program located in the amygdala but a cognitive construction made possible by the cortex (LeDoux, 2016). This means that therapy necessarily involves harnessing and integrating top-down and bottom-up feedback loops. In other words, as visceral sensations emerge we construct new and more adaptive meanings about these sensations. We do so by integrating and balancing our mental operating networks.

THE MIND'S OPERATING NETWORKS

What we call the *mind* is composed of self-organizing mental networks that emerge from subordinate sensations, emotions, feelings, and thoughts. Optimally, our mental networks work together to maintain a coherent sense of self as we balance the needs of our body while we adapt to our environment. The mind keeps the self organized by balancing the feedback loops between the salience, executive, the default-mode networks.

The Mind's Operating Systems

Salience Network

- Referred to as the "sentient self" (the material "me")
- Detecting emotional and reward saliency
- Detecting and orienting toward external events in bottom-up fashion
- Composed of the bilateral anterior insula, dorsal anterior cingulate, and amygdala

Default-Mode Network

- Reflecting, spontaneous thoughts, and mind wandering
- Activated during tasks of mentalizing, projecting oneself into the future or past
- Activated when reflecting on social relationships
- Composed of the medial prefrontal cortex, hippocampus, posterior midline, and cingulate cortex

Central Executive Network:

- Moment-to-moment monitoring of experience (metacognition)
- Responsible for selection, planning, and decision making toward goals
- Working memory that helps select, orient, and maintain an object in the mind
- Composed of the bilateral dorsolateral prefrontal cortex and posterior parietal cortex

These mental networks organize self-referential information derived from long-term memory and current experience. The disorganization or fragmentation between the networks contributes to psychological disorders. Consider the following examples:

- Kyle spent an inordinate amount of time ruminating about previous conversations and unfortunate events. As a result of excessive default-mode network activity, he missed what was going on in the present moment.

- Barbara stewed in the emotions in the moment. She justified her excessive preoccupation with her body sensations and misinterpreted the feelings of others as threatening. She activated her salience network while minimizing activity in the executive and default-mode networks.
- Aaron was a successful computer engineer but not successful in his relationships, never reflecting on them or in touch with his own body and emotions. He had underdeveloped his salience and default-mode networks.

Each of these people worked in therapy to rebalance their mental operating networks.

The Salience Network

The salience network, sometimes called the feeling network, involves somatic sensations and the emergence of emotions. It provides mental operations that "self"-organize somatically derived information. As the name suggests, the salience network selects stimuli that stand out as salient from emotionally irrelevant stimuli competing for our attention. The salience network helps us determine what is relevant to our best interests on an emotional level. Based on visceral and subsequently generated emotional information, the salience network contributes motivation to decisions to act or not to act.

Emotions are constructed from bottom-up sensations and top-down cognitions, which are generated and modulated through brain-body feedback loops (Damasio, 1994; Craig, 2015). The neuroanatomical pathways that carry interoceptive information from the body are read by the insula cortex (*insula* means "island") located deep within the large fold called the Sylvian fissure. The mid-insula contributes to interoceptive awareness, the subjective feeling of being alive as the "material me" (Craig, 2015).

These interoceptive feelings come to awareness in the anterior insula, with the experience of ourself as an emotional being with subjective emotions such as happiness, sadness, anger, disgust, and fear.

Because of its proximity to the prefrontal cortex, the anterior insula generates intuitive visceral (gut) feelings to influence the decisions we make. Only a few other species, including whales, dolphins, elephants, and our close ape cousins, are endowed with a large anterior insula. All these species are considered to have a rich emotional life and relatively complex social systems and thrive from close bonding. This makes the insula a key intersection between the emotional feelings of being an indi-

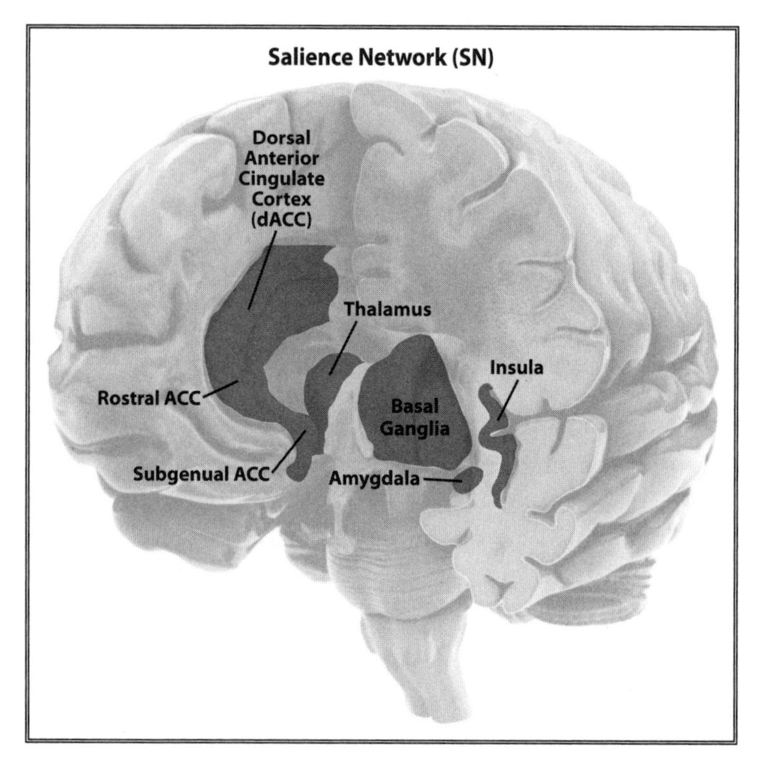

Figure 1.1: The Salience Network as detailed above includes the dorso anterior cingulate cortex (dACC), the rostral anterior cingulate cortext (rostral ACC), subgenual anterior cingulate (subgenual ACC), thalamus, amygdala, basal ganglia, and insula.

vidual and the feelings of being with other individuals. The insula and anterior cingulate cortex are endowed with spindle cells, with long axons that transfer immediate and emotionally compelling insights.

In the examples above, Barbara personified the old proverb, "Too much of a good thing can be a bad thing." She was so overwhelmed with her own body sensations and emotions that she was anxious around other people. Therapy for her involved cultivating the talents of her other networks to be present and engaged with others, and to reflect on those relationships. It also involved enhancing the sensitivity of the salience network to become more socially intuitive.

The subjective experience of pleasure contributes to Barbara's felt sense of self. The mid-insula is directly modulated by the nucleus accumbens, which functions as one of the principle pleasure centers of the brain. With rich connections to the thalamus, amygdala, insula, cingulate, and prefrontal cortex, the nucleus accumbens is well positioned to receive input about potential reward opportunities. When it receives dopamine, the potential opportunities are coded in so that Barbara becomes motivated to achieve the reward. Successful top-down reappraisal enhances emotional regulation of how she labels and approaches an experience. For example, Barbara might say about her social anxiety, "I am excited to meet new people."

Because much of the environment may be ambiguous or irrelevant to her needs, assessing relevance was critical for Barbara. Serving as a relevance detector, the amygdala plays a role in guiding choices in ambiguous and unpredictable situations. Compelling her to act with little attention to the subtleties and nuances, especially when detecting potential threats, Barbara's amygdala may have become overactive. When coupled with overactivating the right insula, she felt at times that the salient threat was coming from her own body and spurred a panic attack. By leaning toward instead of avoiding body sensations, Barbara activated her left insula, which is associated with the alleviation of panic. I pick up this thread in Chapter 7.

The Executive Network

Our prefrontal cortex (PFC) is larger than in any other species, comprising approximately 30 percent of our brain. Our species has evolved the capacity for complex decision making, holding more than one thought in mind, and planning for the future. These enhanced cognitive skills require the ability to regulate our emotions so that immediate gratification does not sabotage long-term gain. The expansion of the PFC made these skills possible through the emergence of the executive network.

More than any other species, we can maintain and enhance top-down control over bottom-up circuits. For example, my family once had a cat that had been abused before we adopted him. In response to low voice tones and abrupt moments, he flinched or ran away. Despite the years of living in our nurturing home, he never grew out of those responses. With a PFC consisting of only 3.5 percent of his brain, he had little to no top-down control over those immediate responses. We, on the other hand, do possess or have the potential to develop those skills through our sophisticated executive network.

Due to its size and complexity, our PFC features areas with a variety of different executive functions. The dorsolateral prefrontal cortex (DLPFC) is the most evolutionary advanced part of the brain and does not fully myelinate until we are in our midtwenties. Executive functions, including attention, problem solving, and working memory, equip the DLPFC with the capacity to provide executive skills that enable us to focus on a task to completion. Working memory is more complex than simple short-term memory, as it provides a significant measure of executive control over what information can be held "in mind" for roughly twenty to thirty seconds. Accordingly, the DLPFC provides the infrastructure to perform goal-directed behaviors by linking together in conscious awareness the complexities of the task at hand so they can be attended to and manipulated. In therapy, these capacities play an indispensable role in regulating follow-through on behaviors that reduce anxiety and depression.

Underdevelopment and underactivity in the executive network can contribute to attentional disorders. Also, a person with DLPFC deficits may develop the paradoxical syndrome known as "pseudo-depression," which is marked by a lack of spontaneity and affect, rather than negative affect. In other words, the person looks depressed but denies depression when asked.

The DLPFC, along with the posterior parietal cortex, which is key in registering where we are in space (like a GPS), is involved in executive decision making about where to move and potential long-term outcomes. The executive network, sometimes referred to as the central executive, operates like the CEO of the mind. I expand on its role in generating belief and attention to the present moment in Chapter 10.

In the examples at the start of this section, Aaron's executive network functioned like a laser. His attention, working memory, and goal-directed behavior brought him accolades at his tech firm. Unfortunately, he was mocked as a Spock-like *Star Trek* character. His coworkers, who generally

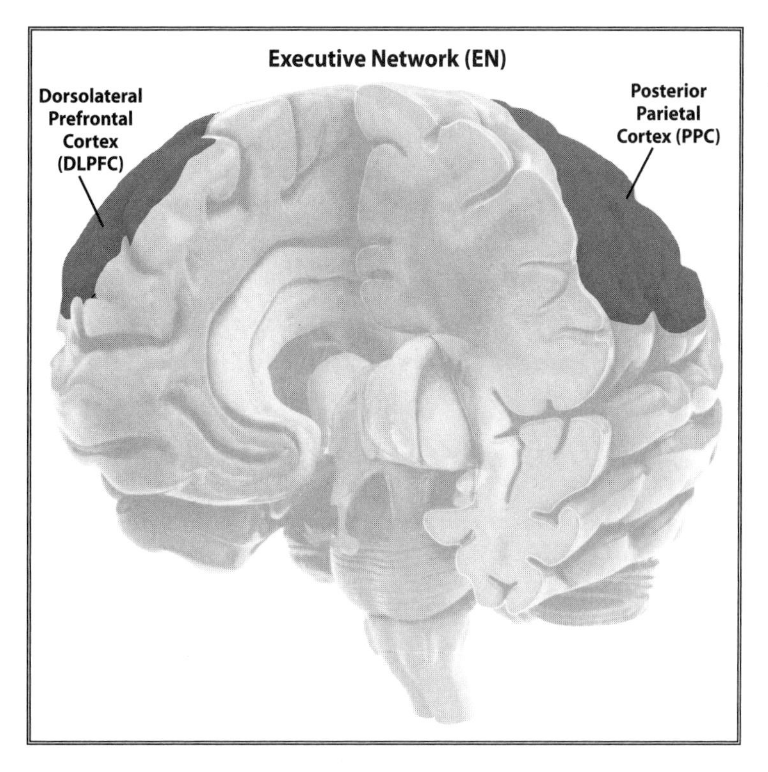

Figure 1.2: The Executive Network as detailed above includes the dorsolateral prefrontal cortex (DLPFC) and the posterior parietal cortex (PPC).

considered him emotionally unavailable and socially inept, preferred to work with him only at arm's length. Therapy with him entailed cultivating his salience network to augment the talents of his executive network. He was also encouraged to reflect on his relationships with his default-mode network.

The Default-Mode Network

In the movie *Analyze This* that I referenced at the beginning of this chapter, Doctor Sobel fades off into daydreaming while Caroline complains about her breakup. In his imagination he jumps up from his chair and shouts, "You are a tragedy queen! 'Steve doesn't like me'?! . . . 'Steve doesn't respect me'?! . . . Get a f—ing life!!" The scene fades into showing him sitting in his chair looking off into space, imagining that scene. Perhaps wondering if he was listening, Caroline says, "Doctor Sobel? Doctor Sobel?" He is shown consciously reentering the room from his fantasies, with his executive network back in control, and then responding by saying, "Oh . . . I was just reflecting on your situation. I will think about it, you will think about it, and we will pick this up at our next session."

Just as Doctor Sobel faded off in the movie, most people spend up to 30 percent of their non-sleep-time daydreaming, ruminating, and simply not engaging in the present moment. During these periods we tend to operate on autopilot, functioning throughout the day performing habitual tasks without our conscious attention, such as driving with "highway hypnosis" or brushing our teeth while reflecting on the day. You may reminisce about last night's dinner or plan the one for tonight, but when a car swerves into your lane your executive and salience networks kick into gear to be attentive to the present moment and the danger it presents.

We all daydream, fantasize, plan for the future, and hold imaginary conversations in our mind. In the field of hypnotherapy they have been referred to as ultradian rhythms (Rossi, 2002). These periods of reverie have only recently been the subject of inquiry in neuroscience. The midline area of the medial prefrontal cortex and the parietal areas are active during times when we are not focused on our immediate environment. The more our focus is self-referential—that is, on our feelings, emotions, and self-related thoughts, including fantasies about the future and ruminations about the past—the more these midline areas are active.

These self-referential reflections are associated with what is referred to as the default-mode network (DMN). When the mind is not engaged in the immediate environment, it falls back to a "resting state," the default mode (Raichle et al., 2006). It draws upon autobiographical memory and

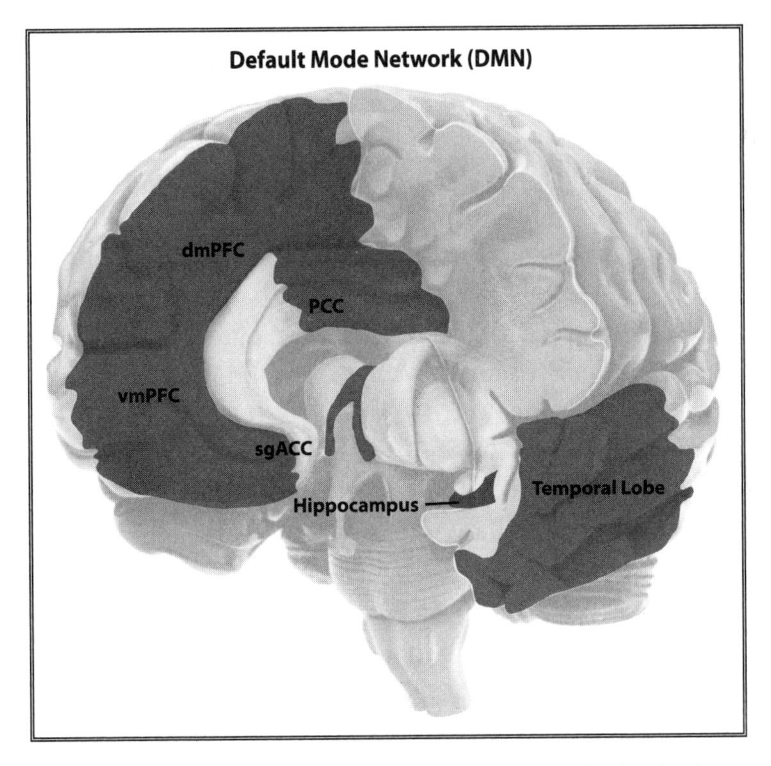

Figure 1.3: The Default Mode Network as detailed above include the the dorsomedial prefrontal cortext (dmPFC), ventral medial prefrontal cortex (vmPFC), subgenial anterior cingulate cortex (sgACC), posterior cingulate cortex (PCC), and the hippocampus, deep with the temporal lobes.

personal judgments. While the DMN is active during the retrieval of episodic memories, it is inhibited during the encoding of them. In other words, when you are remembering something that happened, your DMN is active, but when you are experiencing the event that you will later remember, your executive and/or salience network is active.

> A client can benefit by your inquiries about what he reflects on during quiet moments. When the stories are negativistic, gently nudging him toward a "can do" instead of "can't" storyline can help him use his default-mode network more constructively.

The DMN appears to develop over time, with a coherent network operating by age seven. At this point we begin to develop the ability to have a mental life and reflect on our relationships. While mature reflection parallels the development of the medial PFC, damage to this network can severely compromise self-experience (Northoff & Bempohl, 2004).

The DMN provides flexible self-relevant mental explorations to anticipate, plan, and evaluate future events before they happen (Buckner, Andrews-Hanna, & Schacter, 2008). This network allows us to mentally time travel, playing out how situations may have occurred differently in the past or how they can turn out positively in the future. The DMN is engaged when we are telling ourselves stories about ourselves or about other people. Because activity in the DMN also involves thinking about other people, we play out past or potential conversations by fantasizing about how things could have been different or how we hope relationships may develop.

The fantasies generated in the DMN can boost or undermine self-esteem. A person can ruminate about past unfortunate experiences or imagine positive future ones. Women who incurred adverse childhood experiences have been found to have less functional connectivity within the DMN, and it appears that early trauma interferes with its development (Bluhm et al., 2009). In the examples given above, Kyle's excessive inward orientation with excessive rumination promoted depression, facilitated by an abnormally active DMN (Grimm et al., 2008a, 2008b).

Depression may reflect the inability to inhibit the DMN during external tasks. Consistent with the DMN overactivity hypothesis for depression, discussed in Chapter 9, those who have been treated successfully have been shown to normalize DMN activity (Posner et al., 2013).

Constructing solution-oriented stories is fundamental to mental health. The extent to which we can portray ourselves as someone who can affect the course of our life, as opposed to being the victim of external causes, is fundamental to the narrative of self-agency (Bandura, 1997). Therapy essentially retrained Aaron's DMN storyline. Imaging potential future positive scenarios built his sense of self-esteem to promote healthy "self"-organization.

> Teaching clients to tell themselves coherent realistic and positive stories about their life and repeat them between sessions can be an important part of therapy.

BALANCING THE NETWORKS

Optimally, the three self-referencing mental networks maintain balance and work together to provide the self with dynamic stability. Balance among the self-referencing networks contributes to a person's coherent sense of self. By reflecting on the past and projecting to possible positive futures, we use our DMN to maintain continuity over time and use our executive network and salience network to think or feel in the present moment. The switching between the mind's operating networks depends on the demands of our immediate environment. Facilitating balance among the feedback loops between these networks represents a critical goal of therapy.

During the performance of cognitively demanding tasks the executive network increases in activation while the DMN decreases. However, there can be a dynamic coupling of the two systems, where they work together to produce creativity. Brain activity that is more lateral (outside) tends to be more cognitive and part of the executive network, while reflection that is more medial (inside) tends to be more self-referential, emotional, and more associated with the DMN and salience network. So, in addition to the top-down/bottom-up feedback loops, outside/inside work together. Balance among these networks is critical for mental health.

The anterior insula and anterior cingulate cortex that form much of the salience network provide the capacity to identify what stands out as noticeable, meaningful, and behaviorally relevant. The salience network monitors information from within (internal input) and from the external world and functions as a controller or network switch that decides what information is most urgent. With its insula and anterior cingular cortex components, the salience network identifies relevant sensory information worthy of attention and acts as a central switch, regulating relationships among the DMN and executive network. Such as when a car swerves in your lane, it switches off the DMN so that the executive network can marshal its cognitive and attentional resources for the task at hand.

A person who has been traumatized may find it difficult to encode new, more positive memories. Resting-state connectivity between the DMN and the right amygdala correlates with the presence and severity of posttraumatic stress disorder symptoms (Lanius et al., 2005). The goal of psychotherapy is not to eliminate DMN activity but to limit of how much time a client spends ruminating and disconnected from the immediate environment. This means therapy promotes more engagement in the present moment by the action-oriented executive network. Enhancing attention to recent positive events and engaging in future-oriented thoughts and behaviors can inhibit negative rumination.

The Mental Operating Systems and Therapy

- The salience network: Therapy has to be emotionally relevant, feel "right" and meaningful.
- The default-mode network: Therapy needs to provide new positive narratives to reflect on.
- The executive network: Therapy must aid present-oriented, action-directed, and productive behavior.

Together these mental operating networks construct meaningful patterns that help us adapt to our environment. For clients who suffer from anxiety or depression, the gains from therapy help them balance the activities among the networks so that they are better able to feel safe, develop the capacity to attend to the present moment, reflect positively on experiences, and imagine a hopeful future.

LONG-TERM MEMORY SYSTEMS

So far I have focused on the mind's salience network, executive network, and DMN and how we "self"-organize. By keeping these mental operating systems in balance and flexibly responsive, we can calm ourselves down or lift ourselves back up, while we stay engaged in the world. Those networks emerge and draw self-referential information from cohesive memory systems to provide ongoing feedback to maintain continuity for a sense of a coherent self.

Multiple memory systems form the background fabric of the self and provide a dynamic information bank for the mental operating networks. The memory systems are always in the process of being modified in response to the changing vantage point of present challenges. Ongoing memory modification feeding back to the mental networks helps us update self-referential information to anticipate future events based on what has happen in the past. Unlike working memory, long-term memory systems require neuroplasticity to encode information, making actual structural changes in the brain. Long-term memory is organized in two broad forms: nondeclarative and declarative memory. Nondeclarative memory is also referred to as the implicit system because it provides feedback to what we do and how we feel implicitly without having conscious awareness of it. The salience network derives information from it to generate emotional states. Declarative memory is also referred to as the explicit memory system because it feedbacks into our explicit awareness and can be declared into recall by our executive and default mode networks.

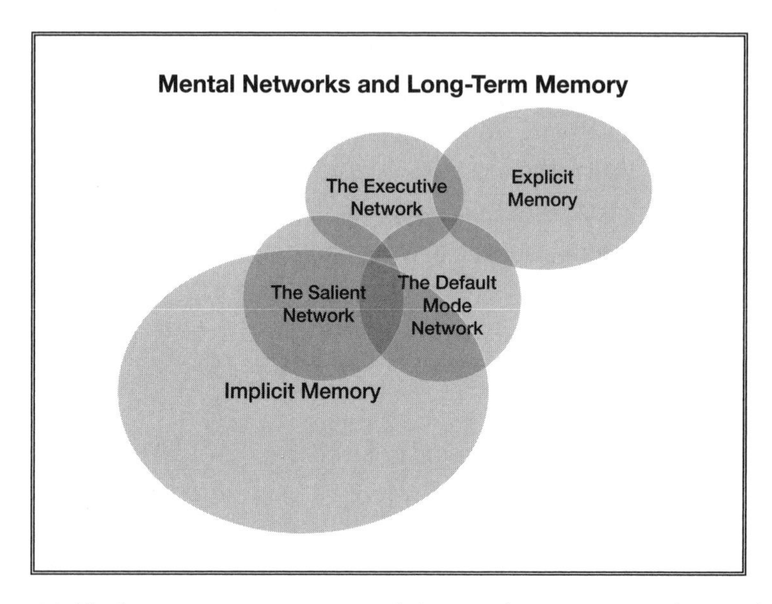

Figure 1.4: The long-term memory systems and the mental operating networks. Note how much larger the implicit memory system is relative to the explicit. Actually, it should be depicted larger than the image above permits.

The Long-Term Memory Systems

Implicit	**Explicit**
Nondeclarative	**Declarative**
Procedural	Episodic
Emotional	Autobiographical
Generalized	Context specific
Classical conditioning	Semantic

The implicit memory system is much broader than the explicit memory system. In other words, what we are aware of about ourselves is only the tip of the iceberg of much larger nonconscious fabric of emotional patterns and habits. The implicit memory systems include procedural and emotional subsystems. Procedural memory involves motor skills coded in our basal ganglia, especially the stratum. These are habits that we learn, such as when we riding a bicycle, type, or drive a car, all of which we do without needing to think about it. These habits represent what we do automatically as if on autopilot; once learned, they are immediately expressed if cued. For the most part they are practical, but sometimes they include habits that have morphed into addictions, as explored in Chapter 6. Much of the implicit memory system includes emotional reactions. Because the implicit memory systems go online first, they influence and set the tone for encoding of explicit memories. When we retrieve autobiography memories, the parameters for those memories are set by implicit emotional memory tones.

The durability of our implicit memories is much greater than for our explicit memories. While the amygdala-based implicit memories are hard to forget, as if written on a stone tablet, the hippocampal-based explicit memories are constantly being modified within the current contextual situation. The acquisition of new habits requires repetition and concerted effort to achieve neuroplasticity initially. For example, when learning a new custom of greeting in the Western world, we fuse implicit, striatum/habit-driven behavior with various semicognitive constructs as we reach out to shake a hand of someone we meet. We respond to his greeting, "How are you?" by replying, "Fine, thanks. How are you?" Neither of us may actually feel fine, but we nevertheless respond automatically with courtesy. We can also read an entire page while daydreaming in our DMN and not remember what we read. Implicit, habit-driven behavior may utilize very little of the executive network, but our DMN and salience network many be quite active.

Just as water flows downhill, so too the brain does what comes easily. This means that the development of new habits requires more input from the executive network initially before the new habit becomes automatic. Once established, the habit comes easy. These habit-based memories will eventually take little attention or energy to occur automatically.

> Because cells that fire together wire together, learning new habits must be repeated to strengthen them, so that they come easy. Habits are strongest if the cues are varied and accessed easily. Clients need to know that it is always harder in the beginning to establish a new,

healthy habit. If they hang in there long enough, their brain will rewire so that the habit will come easy, with little effort, because cells that are wired together will fire together each time.

Procedural memory can provide cues for explicit memory. For example, while driving a car we can listen to the news and comment on what we just heard to a friend in the passenger seat. Later when driving through the same neighborhood we may begin thinking about that same news story. Or we may implicitly feel the anger we felt while driving in that neighborhood but not experience the explicit memory of the news story.

Implicit memories form much of the fabric of our "self"-organization. These varied, intertwining threads include not only procedural memories, composed of movement patterns, such as the way we express ourselves with body language and tone of voice, but also emotional memories, such as reacting with fright to a person speaking with a particular accent. These self-representations include our skilled movements, customs, cravings, and habitual speech patterns, as well as autonomic thoughts, obsessions, and compulsions (Grabiel, 2008). These implicit memories are learned habits, some classically conditioned, repeated, and woven together so that body sensations, movements, and emotions represent each person's sense of individuality. As Caroline told Doctor Sobel, her desire was to find herself as "a person."

Declarative (explicit) memory is recalled through the executive network and DMN. Nondeclarative (implicit) memory represents feedback from subcortical areas, such as the amygdala, basal ganglia, and nucleus accumbens, circuits of the salience network. Accordingly, implicit memories are associated with emotion, movement, habit, and reward and are drawn up by the salience network, which gets reactivated as feelings in the moment.

Explicit semantic memory includes facts about what happened in the past represented symbolically, with words, so that the facts have meaning. Because the mind comprises meaning-making operations, semantic memory contributes to the cognitive skills necessary to facilitate complex meaning. A type of explicit memory referred to as episodic contains self-reference information about prior experience. Episodic memory, dredged up to use during DMN phases, allows us to reexperience events in our lives.

Explicit memory depends on the hippocampus, which serves as a librarian. It codes novel memories and integrates them within circuits in the cortex, which serves as the library. The hippocampus is a novelty detector that compares incoming information to already stored knowledge. Its specialty is binding new to old information.

Whereas nondeclarative implicit memories are often encoded nonconsciously, it may be difficult to recall them later consciously. There is an old joke with some truth to it that, if you lose your keys when you are drunk, it is easier to find them while drunk than while sober. In other words, where you left your keys is coded more in your implicit than your explicit memory. Combined and referred to as state-based memory, the mood you are in when you code an explicit memory represents the same neurocircuits that cue best for recall. State-based memories are quite relevant when working with people recovering from trauma, as we will discuss in Chapter 8.

The mental operating networks derive self-referencial information from the long-term memory systems to provide "self"-organization. The DMN especially draws on long-term explicit memory to reflect on the past and to draw meaning from it. The executive network uses the logical explicit information to make decisions. And the salience network uses the emotional and somatic implicit memories to generate feelings about ourselves. Together, the mental operating networks and implicit and explicit memory help us make decisions about the future.

Implicit forms of emotional regulation can be quite effective (Johnston & Olson, 2015). For example, one study examined how a person integrated incongruent implicit information such as seeing a picture of a happy face overlaid by the word *fear*. He experienced an emotional dissonance that prepared him to exert the control needed to override the conflicting verbal information, producing an "emotional conflict adaptation effect" (Etkin et al., 2006).

When prior knowledge comes into conflict with new knowledge, processing the disparity involves a top-down inhibition of the implicit habitual reactions (Morris, 2007). This means that areas of the prefrontal cortex and hippocampus help compare past knowledge with the new verified knowledge. The anterior cingulate cortex plays a role in conflict resolution by monitoring the person's new behavior compared with old assumptions (Braver, Barch, Gray, Molfese, & Synder, 2001).

THE FAST AND SLOW TRACKS TO THE AMYGDALA

Many people find it confusing when they feel a surge of panic feelings before explicitly understanding the implicit cues for those feelings. Implicit emotional memories are by nature relatively immediate. From an evolutionary perspective the implicit memory system is more ancient than the explicit memory system because threat detection is directly associated with survival. Because the amygdala is a relevance detector, when a threat is present there is nothing more relevant than that threat. The speed of responding could mean life or death.

Accordingly, there are two tracks to activate the amygdala: the fast track and the slow track. In response to an immediate threat in the environment, the fast circuit sends sensory information to the thalamus (the central router of the brain) and then directly to the amygdala. Consequently, a person can sense threat without explicit knowledge or thoughts about that threat. The fast track to the amygdala can trigger the fight-or-flight response and cue flashbacks in posttraumatic stress disorder through this "bottom-up" track. For example, after settling in the United States, Hesham, a Syrian refugee, felt panic when firecrackers boomed on New Year's Eve. He felt what sounded like gunfire before he consciously thought through whether it is actually gunfire. This reaction occurred through his implicit memory fast-track circuit, triggering a flashback.

The slow track to the amygdala in the same scenario may result in Hesham using his executive network to draw up explicit memories and think, "That sounds like gunfire that I heard in my village. But I know it is not." This top-down control allows him to tell himself how to respond based on a reality check, given the context of the situation. If the booming sound is simply a firecracker, he may then wonder why someone set off a firecracker. He may even say, "Oh, yes, it's New Year's Eve, and people celebrate this way!"

The slow track activates the executive network before the subcortical areas of the salience network. The sensory information to the thalamus is relayed to the cortex and hippocampus and finally to his amygdala. Part of the work in therapy with Hesham bolstered his slow track to keep his amygdala activity contained by enhancing the feedback from his executive network. By applying exposure to periodic loud noises, the cued implicit memories combined with new thinking so that his feeling of threat generating a flashback was neutralized by strengthening his slow track in a

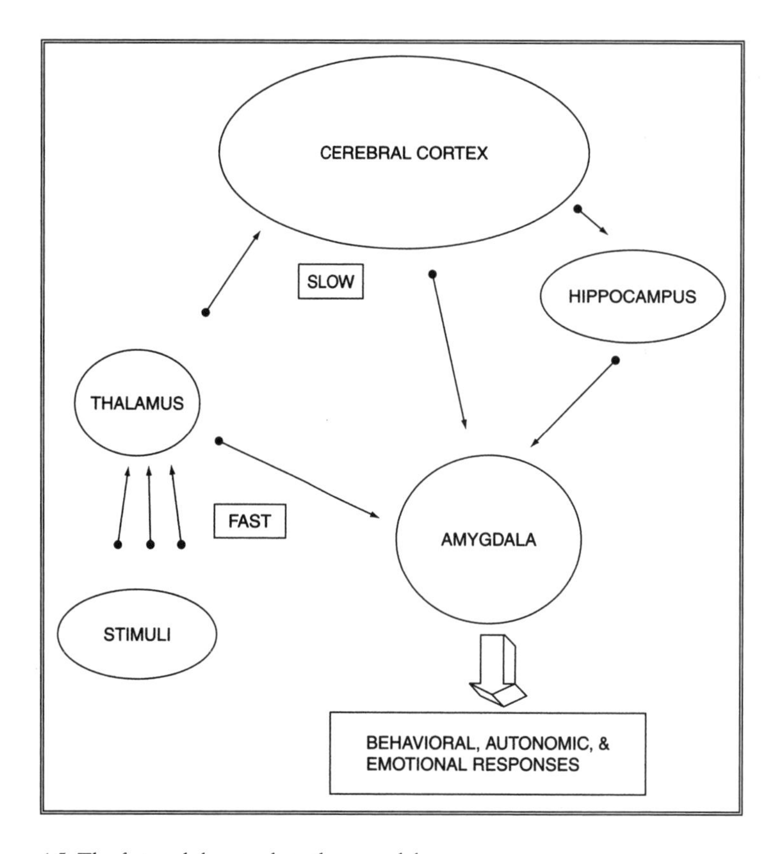

Figure 1.5: The fast and slow track to the amygdala.

top-down process. He dampened the threat-detector functions generated by the fast track by strengthening the connections between his executive and salience networks. Top-down control enhanced his capacity to regulate affect to neutralize threat when there was none. Through this method his previously dysregulated memory systems were reintegrated so that he no longer experienced overwhelming waves of fear that had little to do with the nonthreatening situations.

The slow track can be hijacked by the amygdala so that unrealistic thoughts can emerge fueling fears and phobias. These fears and/or phobias can be attended to in the moment: Hesham may use his executive network to consciously avoid public events such as New Year's and Fourth of July celebrations, or he may use his DMN to ruminate about his fears and phobias. Because explicit memory depends on conscious awareness for encoding, his executive network encompassing focused attention and working memory were needed to form new explicit memories so that Hesham later could "declare" those memories when appropriate. Hesham compared the new the information about celebration to compete with the danger signals he learned previously.

Hesham remembered how planes bombed his village. For him the capacity of subjective time travel was overused and distorted with ruminations about picking through the rubble and finding his uncle dead. During extended DMN periods these episodic memories plagued him. In therapy he learned to use DMN to imagine and project ahead to a positive future. Episodic memory allowed him the capacity to see himself with a past and project ahead to a potential future.

Maximizing Hesham's executive network attentional skills and working memory as well as salience network skills of emotional engagement were critical to transfer information gained in therapy into long-term memory. Because his amygdala is a relevancy detector and his hippocampus is a novelty detector, by maximizing both emotional engagement and novelty, new safety associations were co-constructed in therapy.

Because therapy integrated his long-lingering implicit memories with newly constructed explicit memories, the context of what happened in the past in his village could be reconsolidated within a newly constructed sense of self-efficacy in the present. Therapy simultaneously promoted greater self-knowledge and affect regulation. This was achieved by reconsolidating and integrating the memory systems to match the retrieval cues that promoted self-efficacy.

And what about the questions posed at the beginning of this chapter about a central "me" of individuality? Hesham's identity was challenged and then transformed. Initially, his implicit memory cues complicated his adjustment to a safer community. Through therapy the balance among the feedback loops between his executive network, salience network, and DMN involved progressive nonlinear leaps to higher levels of organization of affect regulation, insight, and memory. "Self"-organization generated by his integrated and reconsolidated memory systems transformed his thoughts and feelings about what he had previously experienced. In short, Hesham's self-image transformed from of a casualty of war to that of a durable survivor.

Therapy involved building a cohesive and positive model of himself by facilitating self-exploration and self-reference, so he could identify, label, and accept feelings and practical needs. His growth and development as an individual included a positive self-reflective internal life within the DMN, the feelings of coherence of the salience network, and productive self-determination directed by his executive network.

For mental health the mind's operating networks—the salience network, executive network, and default-mode network—need to be in balance so that the feedback loops between them work toward enhanced "self"-organization. As the following chapters illustrate, mental health can be undermined and dysregulated by insecure attachment, neglect, and abuse, leading to gene expression or suppression, changes in immune system regulation, and poor lifestyle practices. When the mind's operating networks become out of balance and unintegrated, psychological disorders emerge. Mental and physical health depends on the therapeutic rebalancing of the mind's operating networks and the integration of the long-term memory systems.

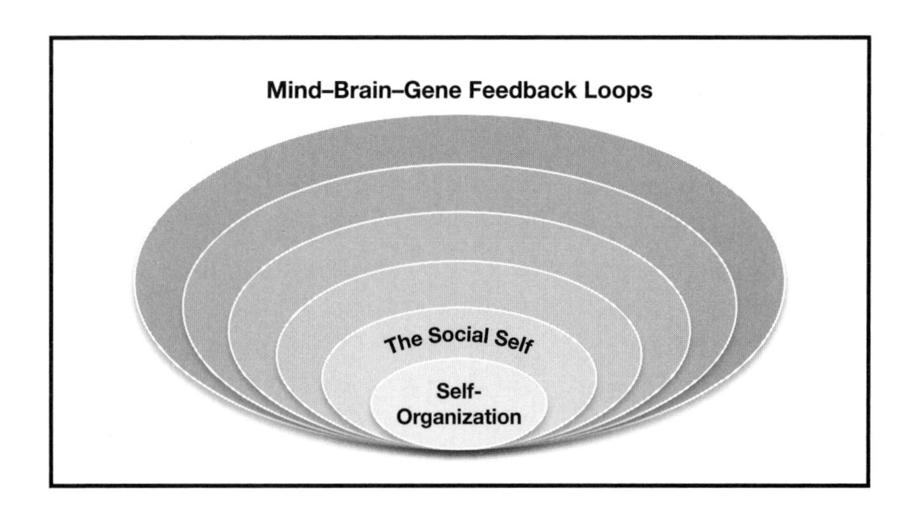
Mind–Brain–Gene Feedback Loops
The Social Self
Self-
Organization

TWO

The Social Self

> For those communities, which included the greatest number of the most sympathetic members, would flourish best, and rear the greatest number of offspring.
>
> —Charles Darwin, *The Descent of Man*

During the first few minutes of her first session Clara complained, "My parents never really wanted me." She described how both parents were highly successful professionals who had little if any time for her. With grandparents far away and no siblings, she had no sense of family. To please her parents she became one of the top students in all of her classes through graduate school, yet that did not garner their attention. She complained that she felt empty, adding, "There's always been this feeling that there is something missing in me."

As an adult she attempted to establish a close intimate relationships, but her brief romantic encounters ended with both her and her potential partner acknowledging the existence of an emptiness between them. She asked me, "Where does this emptiness come from?" With her aging parents needing her help to move into an assisted living environment, even the thought of spending time with them was annoying. In response to this expectation, she hired a social worker not only to deal with the move but also to talk to them about their "incessant complaining about losing their independence."

She looked at me with astonishment and asked, "What am I supposed to do with their feelings?" Almost in the same breath she noted that she asked the same question to the man whom she dated for the third and last time. "All he could say was, 'If you don't know, I can't help you'!"

Humans are fundamentally a social species. We thrive within positive relationships and suffer without them. Clara complained that she felt that there was something missing in her. The implicit emptiness she felt with others mirrored her underdeveloped sense of self. Though she managed to excel through graduate school with a well-developed executive network, her salience network was impoverished and used her default-mode network to bemoan relationships gone sour.

How malleable are our brains to fill the emptiness that she described with implicit memories of insecurity? Are there circuits in the brain that would enable her to develop the capacity for secure relationships and even intimacy? If so, how might psychotherapy promote this capacity? This chapter answers these questions. It is useful to begin by exploring how early experiences shape an infant's brain.

NURTURED NATURE

The brain of infants has considerable redundancies and many more neurons than that of the obstetricians who deliver them. There is a significant degree of extra undifferentiated potential that permits them to adapt to their unique family, community, and culture. The infant's brain is also far more malleable than an adult's. Of all other species, humans spend the longest time in the hands and supervision of caregivers. Our brain is wired by experience in a social context, and the range of possible social experiences is vast.

During the first year of life the right hemisphere, relative to the left, tends to be more active (Chiron et al., 1997). The right hemisphere is more adept at picking up the nonverbal emotional nuances of communication so dominant in infant-parent communication. Early development

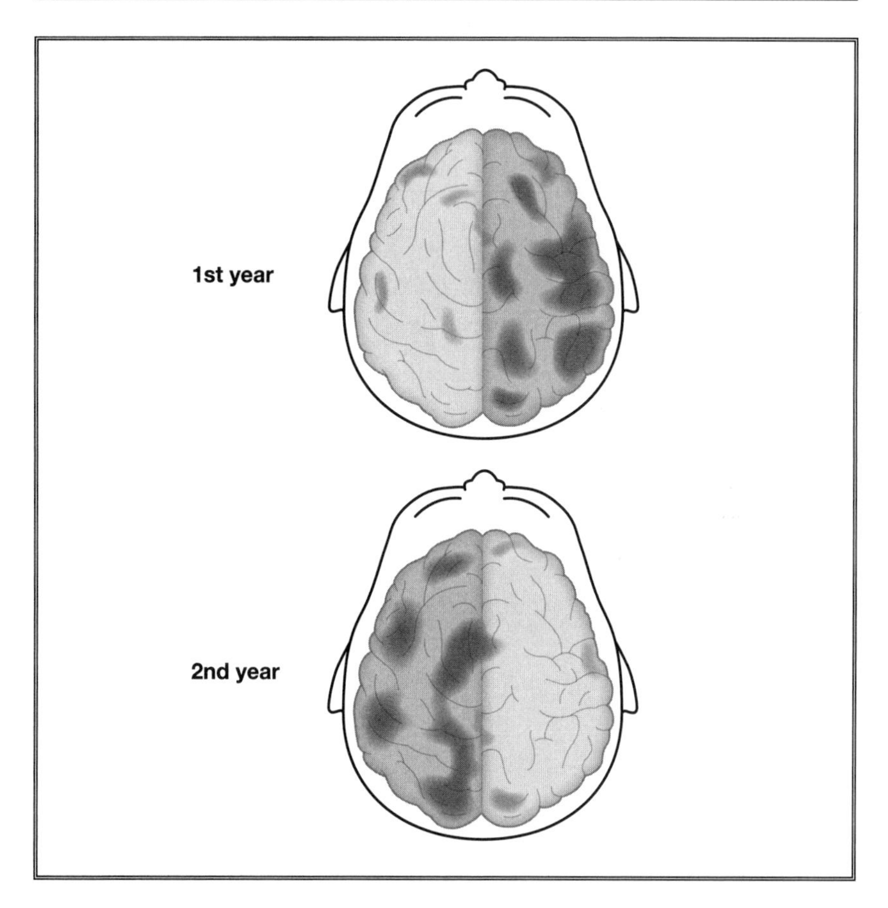

Figure 2.1: The relative dominance of each hemisphere as a child develops. During the first year of life, for example, there is relatively more activity in the right hemisphere followed by the second year with more activity in the left hemisphere.

builds implicit-emotional memory of whether or not approaching people offer comfort and safety. For Clara avoidant attachment became her implicit memory system and mode of social interaction. As she grew older the socially sanctioned reference points for intimate relationships collided with her discomfort and lack of intimacy-building skills.

SOCIAL BRAIN NETWORKS

Though underdeveloped, Clara's social brain circuits needed to be primed. Though there is no actual "social brain," some brain circuits are associated with social interactions. Psychotherapy is the interpersonal sculpting that integrates these circuits (Cozolino, 2017), which include the orbital frontal cortex, insula, cingulate cortex, and a few socially sensitive types of neurons. The anterior insula and anterior cingulate contain a unique type of neuron, known for its spindly shape, unique size, and highly connected quality, called the spindle cell.

Social Brain Components

- Orbital frontal cortex
- Amygdala
- Insula
- Cingulate
- Temporal parietal junction
- Mirror neurons
- Spindle cells
- Facial expression modules

Highly social mammals who have been observed displaying empathy-like behavior, such as whales, elephants, and great apes, also have well-developed anterior cingulate and anterior insula (parts of the salience network), and spindle cells within those structures (Allman et al., 2011). At birth humans have approximately 28,000 spindle cells, growing to 184,000 by age four and 193,000 by adulthood. By comparison an adult ape has 7,000.

Spindle cells are rich with serotonin circuitry, intimately connected to mood, social ties, expectation, and reward. They provide a means to grasp an intuitive sense of the emotions of another person, enabling us to make snap, intuitive judgments that are fused with emotion and social sensitivity. While people like Clara need to learn to kindle their activity, spindle cells are vulnerable to abuse and neglect, and they may be impaired in neurodevelopmental disorders (Allman, Watson, Tetreault, & Hakeem, 2005). We can hypothesize that therapists routinely cultivate spindle cell activity to hone alliance-building skills and to detect the emotional nuances of their clients.

So-called mirror neurons were initially discovered in macaque monkeys when they fired after the monkey observed the movement of another individual (Rizzolati & Arbib, 1998). For example, when an individual observes another move an arm but does not do so himself, he fires neurons as if he is moving his own arm. It is believed that mirror neurons were useful during evolution to predict goal-directed behaviors of other individuals. They may have provided an adaptive advantage to predict how to stay safe from potential danger if an observed individual means harm.

Mirror neurons may be found in various areas of the prefrontal cortex, posterior parietal lobe, superior temporal sulcus, insula, and cingulate cortex, which are all associated with social skills. Though they are quite controversial, people have speculated that mirror neurons provide the capacity to experience empathy (Iacoboni, 2004). Others are not convinced that they represent a specific type of neuron. They may be related to motor neurons that are co-opted into firing after observing ourselves move and then later observing another individual make the same movement (Hickcok, 2014). However mirror neurons are understood in the years to come, they and other socially responsive components of the salience network facilitate emotional attunement and empathy. Though Claire had not yet developed these skills, she would cultivate them.

CULTIVATING EMPATHY

The capacity to emphasize with another person is a foundational social skill. As psychotherapy is fundamentally built on social skills, it involves the capacity of the therapist and client to coactivate parts of their brains that bring them closer to a shared understanding. Together these social brain circuits represent the means through which empathy is given and felt.

The area of the brain associated with physical pain overlaps social pain (Lieberman, 2013). Empathy is associated with activity in both the anterior insula and anterior cingulate cortex (ACC) (Craig, 2015). Both of these areas are associated with the anticipation of pain, emotional reactions to pain, and understanding the feelings of others (Singer et al., 2004).

While the ACC and insula become active when we experience pain and when we empathetically witness pain in another person, the somatosensory cortex gets activated when we directly experience pain. The density of oxytocin receptors in the ACC is associated with the degree of empathy.

The salience network, which includes the insula, especially on the right side, promotes gut or intuitive feelings about another person. Together with the ACC, the anterior insula provides the capacity for empathy, which comprises the awareness of our own bodily state. In other words, feedback loops from our body are intertwined with feelings about another person.

People like Clara are likely undeveloped in these areas or have actively inhibited these abilities. As a result, they possess less social sensitivity, blunted empathy, and less awareness of their own feelings and bodily states. At the beginning of therapy Clara seemed ill aware of her own body states. In the extreme, those with severe abuse may also exhibit lower heart rate, arousal, and amygdala activity (Vogt & Sikes, 2009).

It was initially difficult for Clara to develop empathy for her aging parents. An opportunity presented itself during one session, when she was suffering from pain from a strained back and began to express concern for her parents. Through the overlapping neural networks between her physical pain and social pain we were able to bridge discussing her thoughts and feelings about cultivating more empathy for her parents.

Feeling rejected during childhood is the antithesis of empathy. Bullying, one type of social pain, has received considerable attention in the media due to a number of widely publicized adolescent suicides. The core of the emotional social brain, the amygdala and anterior cingulate, is activated by verbal abuse and ridicule and deactivated by social support (Coan, Schaefer, & Davidson, 2006; Teicher, Samson, Polcari, & McGreenery, 2006). Social support can coregulate a child's emotions to promote less amygdala activation and more prefrontal activation (Tottenham, 2014). Buffering social stress with support increases amygdala and prefrontal cortex connectivity to promote secure attachment and reduce anxiety.

THE VAGAL BRAKE

Through primate evolution, the emergence of the systems of the vagus nerve made possible more sophisticated social engagement skills that combined the ability to regulate affect while simultaneously negotiating the subtle nuances of communication. The myelinated vagus nerve system forms the key part of the parasympathetic nervous system. The vagus (meaning wanderer) enervates many of the organs in the thoracic cavity, as well as the muscles in the lower part of the face and vocal cords. The polyvagal system has also been referred to as the social engagement system because people with a well-functioning vagus possess the capacity to regulate their heartbeat down while expressing emotions without defensiveness (Porges, 2011). To facilitate social engagement, the upper vagal system connects with the smile muscles, the inner ear, and the vocal apparatus used in prosody (Porges, 2011).

Social competence necessitates putting the brakes on a rapid heart rate and masking a frightened or angry facial expression when engaging in potentially tense communication. However, people who have endured adverse childhood experiences tend to have an undeveloped vagus system and are more apt to overreact or remain hypervigilant in response to socially stressful situations. The ability to stay calm while engaging others was not a skill Clara's parents taught and so represented a goal in therapy.

The vagal brake, with its social engagement system, and affect regulation are interdependent capacities. Together, affect regulation and positive relationships represent two related aspects of mental health. What happens if a child is exposed to no social engagement? This can happen with maternal depression.

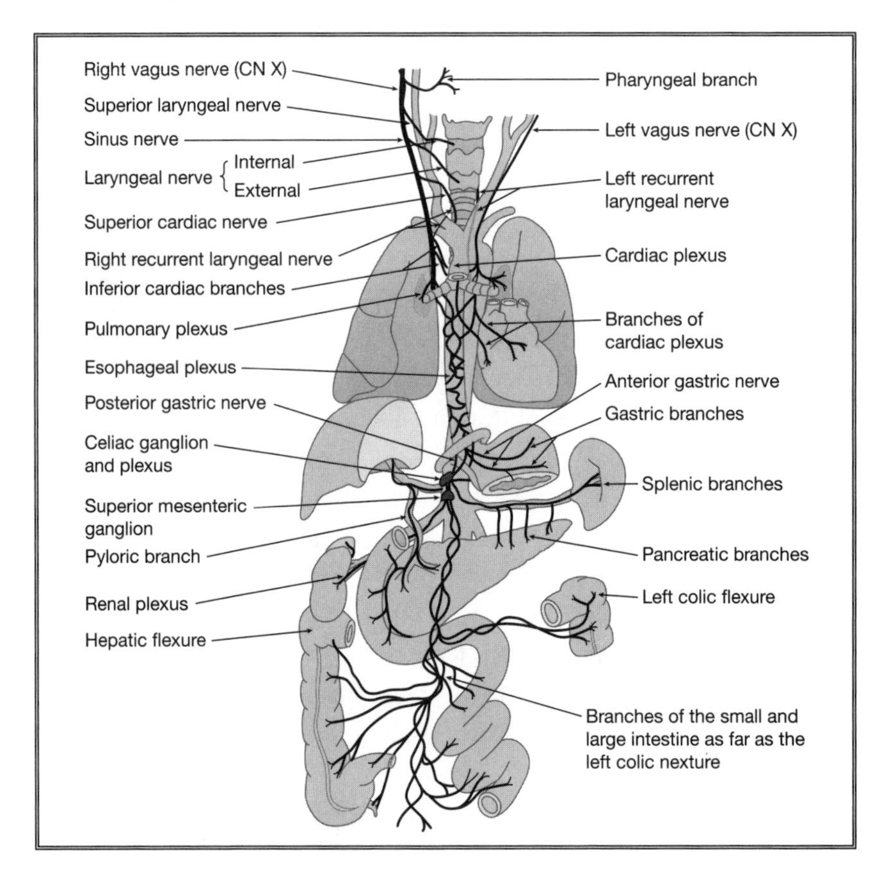

Figure 2.2: The vagus nerve system and its connections to the organs in the thorastic cavity.

The Psychoneuroimmunology of Social Support

- Decreased cardiovascular reactivity (Lepore et al., 1997)
- Decreased blood pressure (Spitzer et al., 1992)
- Decreased cortisol levels (Kiecolt-Glaser et al., 1984)

- Decreased vulnerability to catching a cold (Cohen et al., 1997)
- Decreased anxiety (Coan, Schaefer, H. S., & Davidson, R. J. (2006)
- Increased natural killer cells (Kiecolt-Glaser et al., 1984)
- Slower cognitive decline (Bassuk et al., 1999)
- Improved sleep (Cohen, 2004)
- Lessened depression (Russell & Cutrona, 1991)

THE EFFECTS OF MATERNAL DEPRESSION

Mary, a thirty-two-year-old accountant, was first seen by a psychiatrist in our department for postpartum depression and immediately put on a selective serotonin reuptake inhibitor. During a team meeting I expressed concern that she was receiving only medication and took her on as a client. Mary's depression not only was a problem for her but also was potentially devastating to her infant. For this reason she was asked to come to the first session with her baby.

Like all infants, her baby was primed to engage her, without which he would not thrive. After birth, Mary's son, Austin, reflexively oriented to her eyes, but he did not find warmth and nurturing. He found a blank stare. I immediately engaged him with positive facial expressions, a high voice tone, and animated prosody. I told Mary that, though she did not feel that she had it in her to engage with her son, his brain would thrive with engagement. To demonstrate, I began to hide my face behind my hands, peeking and then hiding again. Austin began to smile, as if awaken from a stupor. Mary did engage him, and to her astonishment, when he smiled she felt a little better each time. It became an antidepressant feedback loop.

The still face paradigm illustrates a snapshot of the devastating effects of parental depression on a child. In the still face paradigm a mother is asked to show no facial expression (Tronick, 2007). In response, the baby initially tries to entice her into playfully interacting. Eventually the baby becomes agitated, distressed, and finally gives up, looking despondent. If exposed to a still face for an extended period, the baby no longer approaches novel toys and receptivity and interactions with the environment shut down.

Parental depression acts to shut down the infant's social brain circuits to neutralize what normally thrives on playful interaction. Because an infant's brain has yet to develop cortical circuits, it relies on subcortical areas such as the amygdala. Because the amygdala is a relevance detector, it cannot tolerate the ambiguity of a still face. Neutral faces provoke more amygdala activity in children than in adults (Tottenham, Hare, & Casey, 2009). As infants mature they learn to tolerate neutral faces and ambiguity by increased cortical processing, especially in the prefrontal cortex (Casey et al., 2005). A fully operational prefrontal cortex helps us navigate through an often ambiguous world.

> If you respond to a client with the psychoanalytically traditional "neutral" stance, presuming that he will develop transference, you may replicate his depressed parent.

We can derive significant implications from the still face paradigm for the long-term mental health of a child and into adulthood. If a child's principle caregiver maintains an expressionless face for extended periods of time, as would have occurred with Mary's postpartum depression, the child is deprived of critical emotional nourishment. Just as a balanced diet, sleep, and exercise are foundational for a healthy brain, so are animated interactions and facial expressions from parents. Positive or negative care can either turn on or off genes critical for the structure of the brain associated with affect regulation and social skills (Callaghan, Sullivan, Howell, & Tottenham, 2014). Failure to thrive because of lack of emotional nurturance can result in a lack of coherence in the salience network.

Mothers with postpartum depression tend to pull back from their babies. They tend to touch their bodies less as early as the second day after delivery (Ferber, Feldman, & Makhoul, 2008)). In response, the child's essential needs are starved by the depressed mother's flat affect, flat voice, neutral expression, and overall dulled interactions. Correspondingly, infants of depressed mothers whose speech prosody is flat show EEG patterns consistent with depression (Wen, et al, 2017)).

> Infants of depressed mothers
>
> - Have overactive right frontal lobes
> - Have underactive left frontal lobes
> - Have lower levels of dopamine and serotonin
> - Display more aversion and helplessness, and vocalize less
> - Have higher heart rates, decreased vagal tone, and more developmental delays at twelve months of age
> - Have higher levels of stress hormones (Field, Johnston, Gati, Menon, & Everling, 2008)

> Maternal depression during the first two years of a child's life is the best predictor of abnormal cortisol production in children at age seven (Ashman et al., 2002).

Infants with a depressed mother are faced with a difficult predicament. They are immersed in a feedback loop in which there is only one way interact: to be sad together so that they can establish coherence to their relationship. Deprived of enhancing "self"organization through mutual growth with their mother, they may take on the mother's depressive mood, irritability, and even hostile affect. When they encounter other people the only way they are able to expand in complexity and self-coherence is by establishing relationships around depressive features that were established with their mother (Tronick, 2007). They tend to develop implicit patterns of coping that entail avoiding others and inadequate attempts to self-comfort, resulting in negative affect and less engagement with the environment.

Because we are complex adaptive systems and by are nature open systems, maternal depression shuts down children's attempts to thrive, stunting not only mood and behavior but also cognitive development. Mutually experienced depression constricts, closes down, rigidifies, and ultimately threatens both the mother and child, with no room to grow.

The mother loses coherence and complexity while her child loses the opportunity to develop the full range of capacities to adapt to a complex interpersonal world.

As an open system, a child must acquire social input from others to maintain and increase coherence, complexity, and distance from entropy. Neither fixed nor chaotic, the child's developing mental networks construct and thrive on adaptive meaningful experiences. Because the child's development of viable mental operating networks is a self-organizing emergent process, all his interpersonal relationships contribute to coherent and unique implicit relational knowledge.

As an infant, toddler, child, adolescent, and adult, the individual progressively leaps to higher levels of organization while feeding on contextual interactions within and between each dimension. The family affects nutrition, physical activity, and attachment style, and they reside in a particular society with a particular socioeconomic and education level and within a cultural context. All these factors contribute to the individual's "self"-organization.

GOOD-ENOUGH PARENTING

Whereas no attunement, such as with maternal depression, is devastating, hyperattunement also undermines psychological development. Half a century ago Donald Winocott pointed out that moderate matching was superior to perfect matching. Essentially, perfect is not good enough. Moderate matching is "good enough." By this he meant that generally consistent but not perfect caretaking builds frustration tolerance for the child.

"Good-enough" parents could have prepared Clara for the often ambiguous, sometimes stressful life experiences that she later encountered. Instead, her parents offered low levels of affective matching, which spawned her insecure attachment. Yet, had she been offered the other extreme, instantaneous soothing, she may have had difficulty developing the vagal brakes necessary to activate her parasympathetic nervous system to counterbalance the sympathetic branch during or after stressful experiences.

Children who receive excessive attention and coddling become spoiled and narcissistic. As adults they may have a tendency to become pessimistic and passive-aggressive, having been trained by their parents that they can be gratified for no effort or even by simply whining. This passive behavior tends to overactivate their right hemisphere, promoting withdrawal and negative emotions, and underactivate of the left hemisphere with its associated positive emotions and proactive behaviors.

Good-enough parenting would have offered Clara moderate matching to bolster the resiliency of her autonomic nervous system, allowing her to rise to the challenge of stressful situations and then calm down afterward to recoup. Good-enough parenting would have allotted time before she was soothed, during which she could have developed the capacity to self-sooth by activating her parasympathetic nervous system to calm herself down.

The quality of the reciprocity within the parent-infant dyad—synchrony, matching, coherence, and attunement—offers the emotional nourishment critical for healthy development. The rhythm and timing of the interaction between mother and infant are key for the developing child to build the capacity to negotiate the nuances and subtleties of relationships. Sensitivity in the midrange is most predictive of secure attachment (Jaffe, Beebe, Feldstein, Crown, & Jasnow, 2001). When bonding/attachment is enhanced by repairing mismatches, healthy "self"-organization progresses. Just as Winnocott hypothesized, good-enough parenting offers flexible matching and reparation. Through encountering many moderate, well-coped-with stressors, the child best develops approach behavior, stress tolerance, and resilience. But when repair does not occur or there are repeated unsuccessful attempts to repair the ruptures, the child develops defensive and avoidant coping behaviors that undermines later relationships. These interactions for approach or avoidance correspond to the ability to differentiate between pleasurable and painful experiences (Berridge & Kringelbach, 2008).

> A good-enough therapist is better than one who pretends to be perfect. Your imperfections model the real world. It is how you use your imperfections that determines the quality of therapy. Using the misunderstandings, conflicts, and points of tension in the relationship as the focus of resolution builds positive outcomes. Mismatches are a plus when resolved. Each self-organized transition, punctuated by emotional coherence and insight, may result in dopamine release from the ventral tegmental area of both therapist and client.
>
> "Good-enough therapy" with moderate matching helps build frustration tolerance and the ability to self-sooth. The "impingements"—affective mismatches and misunderstandings—so inherent to communication represent the grist to be resolved in therapy. How a client may be angry with you offers a valuable opportunity not only to resolve but to leap to a higher level of shared meaning, promoting "self"-organization.

Either high- or low-sensitivity reparation may promote insecure attachment. Low-sensitivity reparations can occur during long-duration mismatches when infants are constantly overwhelmed with stress and so fail to develop coping skills. High sensitivity involves intense vigilance, where matching lasts too long, with little opportunity for reparation. Infants not given a chance to confront sufficient amounts of stress with support fail to develop effective methods of coping (Tronick, 2007). Through moderate matching the infant better develops coping skills and resiliency. The mother has an opportunity to find the midrange of comfort for herself. On the other hand, high or low matching forces the mother in one extreme or the other: blunting her affect or becoming too stressful.

Given that fathers excite and set limits, the research paradigm referred to as the Risky Situation (RS) serves as a measure of the father and child activation relationship. The RS may better measure attachment and affect regulation for boys than the Infant Strange Situation, which may be more appropriate for girls. In the RS, the father's caregiving behavior tends to focus on arousal and excitement as well limit setting (Paquette & Bigras, 2010). In the Rough-and-Tumble Play (RTP) paradigm, the father activates but also sets limits. When fathers do not exercise dominance in the RTP with preschoolers they are more likely five years later to have poor emotional control and high levels of physical aggression, especially boys (Flanders, et al, 2009). RTP trains the child's prefrontal cortex to learn to set limits on affect.

The reparation of the "messiness," rather than synchrony, is key to change in therapy and development (Tronick, 2007). Inherent to complex adaptive systems, we thrive as open, curious people who strive to resolve misunderstanding, and this is a messy process. Through self-organizing interactions we become greater than we were before. Just as the parent-child dyad is a self-organizing system, so too is the therapist-client relationship. Therapist and client adjust to one another in a mutually regulated process. Each of us identifies and makes use of increasing amounts of meaningful information emergent from higher levels of shared understanding achieved together.

SELF-REGULATION AND ATTACHMENT STYLES

Why was Clara so emotionally tone deaf when it came to relationships? She did not encounter adverse childhood experiences or even maternal depression. How does someone like her come to feel so empty? Through preverbal interactions, an infant detects the mother's subtle changes in affect through her voice pitch, prosody, and inflection and through the quality of touch, holding, and facial expressions. Nurturing parents tend to exaggerate joyful and encouraging facial expressions. Their babies learn to detect subtle changes in mood, perhaps even microexpressions (Beebe et al., 2010). These interactions build an implicit memory of attachment or, as John Bolby called it, the internal working model of how relationships work. Accordingly, the mother-infant dyad can be referred to as shared implicit relationship (Tronick, 2007).

Overlapping terms for *attachment*:

- Boding
- Internal working model of attachment style
- Pro-narrative envelopes
- Schemas of being within
- Themes of organization
- Relational scripts

The early patterns of adaptation have been referred to by a variety of overlapping terms. Whatever terminology you prefer, the bottom line for Clara was that the emotional neglect shaped her pattern of "self"-organizing thoughts, emotions, and behaviors. These impoverished interactions formed her self-identity and undermined her self-esteem, depriving her of a sense of lovability. Had she enjoyed a secure attachment experience, she would have been better equipped for durable affect regulation required for intimacy later in life. Her insecure attachment schema led instead to fear modulation and avoidance behaviors, which gave her little opportunity to cultivate self-esteem and social competence.

Attachments styles represent the self-regulation methods people use to cope with stress. While those who are securely attached use multiple strategies for dealing with stress, insecurely attached people are limited to just a few. As a result, their capacity to cope with interpersonal stress can be severely limited. Clara's socially avoidant coping methods were compensated by approach behaviors, such as getting good grades. The attention that she received from teachers and from peers during class projects gave her a sense of a narrow range of belonging. Though positive, it restricted her self-esteem and competence to particular contexts.

The Neurochemistry of Attachment

- Early attachment experiences regulate the opioid system so that the developing child can generate feelings of comfort and buffer stress. Endogenous opioids also play a significant role in providing physical pain relief, and they are also associated with reducing social pain and enhancing pleasure (Curley, 2011).

- Endogenous opioids play a significant role in attachment as part of the infrastructure of feeling soothed with comfort. In fact, variation at one of the opioid receptor genes influences attachment behavior (Barr et al., 2008).
- Feeling soothed with less pain occurs in part through opioid release into the anterior cingulate cortex.
- Maternal nurturance stimulates the release of brain-derived neurotrophic factor (BDNF).
- BDNF and *N*-methyl-d-aspartate expression and increased cholinergic innervation of the hippocampus enhance cognition (Liu et al., 2002).
- Higher levels of BDNF buffer cortisol in the hippocampus resulting from stress and promote ongoing plasticity (Redicki et al., 2005).
- A study comparing the brains of suicide victims with normal controls found lower mRNA levels of the genes for BDNF and receptor tyrosine kinase B (Dwivedi et al., 2003).
- Oxytocin release activates the vagus nerve and so the parasympathetic nervous system, which decreases heart rate and increases relaxation.

The interaction between social support and stress reduction occurs on many levels. The neurohormone oxytocin is a key regulator of emotional and prosocial behaviors (Neuman, 2008). Released by the hypothalamus during close interpersonal contact, it promotes trust and safety. There are many oxytocin receptors throughout subcortical areas, including the amygdala and periaqueductal gray, which when activated inhibit the stress response. Oxytocin receptors on the vagus nerve system activate the parasympathetic system, leading to stress reduction (Gimpli & Fahrenholz, 2001). Because lowering defensiveness is critical for interactions with others, oxytocin facilitates a calming effect along with the capacity of what some have referred to as "mind reading" (Domes, Heinrichs, Berger, & Herpertz, 2007).

Oxytocin has received much attention in both the mainstream press and neuroscience research. A variety of studies have shown that oxytocin appears to facilitate social interactions as well as genetic variations in the levels of stress reactivity, empathy, optimism, and self-esteem, corresponding to physical and mental health (Kumsta & Heinrichs, 2013). Researchers used nasal spray containing oxytocin with subjects and found increased levels of trust and generosity when playing economic games (Zak, Stanton, & Ahmadi, 2007).

SEPARATION, LOSS, AND BUILDING TRUST

Benny and Habib were two five-year-old boys. One had been separated from his mother, and the other lost his in a bombing. The care that they received afterward contributed to dramatically different outcomes. Benny was taken away from his parents due to his mother's drug abuse and inability to care for him. Fortunately for Benny, his mother had reframed from drugs during the pregnancy, with the support of Planned Parenthood. But within a week postpartum depression led to her slip back into drug use, from which she never recovered. Benny was sent to one foster home after another. By age seven he was briefly placed with his grandparents' custody. As his only family contacts, they began the process to legally adopt him, but when his grandmother began treatment for cancer they placed him back in yet another foster home. The visits from his grandparents faded as his grandmother entered a hospice program. Despondent and increasingly withdrawn, Benny's first-grade teacher requested an immediate individualized education program meeting, to include staff from our mental health department and his latest foster parents. This meeting was meant to serve as a pivotal point that brought together a therapy team. Family therapy presented infrequent opportunities for his foster parents to learn to practice building trust and being a social buffer against all of the pain, abandonment, and mistrust (Baylin & Hughes, 2016). However, his foster parents had five other foster children, which provided a major source of their income. Both worked minimum-wage jobs, and they had little time to spend in therapy and were hardly engaged with any of their foster children.

Maternal separation leads to the following:

- The potential use of alcohol and other drugs (Francis & Kuhar, 2008)
- Alterations in the development of inhibitory neurons and changes in the connections of serotonin and dopamine neurons in the medial prefrontal cortex (Helmeke, Ovtscharoff, Poeggel, & Braun, 2008)
- The downregulation of gene expression for GABA (gamma-aminobutyric acid) receptors in the locus ceruleus, resulting in more norepinephrine (Caldji, Diorio, & Meaney, 2003)
- An upregulation of the gene regulating glutamate receptors (Weaver et al., 2016), which contributes to anxiety and depression and difficulty inhibiting negative affect.
- Epigenetic changes to the developing child's stress response system
- Abnormally programmed gene expression in regions such as the amygdala, hippocampus, and prefrontal cortex (Moriceau, Shionoya, Jakubs, & Sullivan, 2009), priming the stress system to become vulnerable to turning on too often for the wrong reason
- Plasticity between the prefrontal cortex and amygdala skewing increasingly toward the amygdala and the rest of the stress system (McEwen & Morrison, 2013)

Habib's situation was dramatically different in terms of support and culture. I met him during a trip to the Middle East as part of a group of mental health professionals training aid workers helping Syrian refugees. His back story was not only traumatic but horrific. One day in Damascus he and his friends noticed a metal object in an alley where they were playing. One of them hit the object with a stick and the bomb detonated. Everyone but Habib died instantly. He lost both legs, an arm, and an eye. Weeks after the incident more bombing killed both parents and his brother. After his hospital was bombed too, he was carried to a refugee camp at the Jordanian border by his only remaining relative. Then his uncle died a few days later. By the time I met him he was in the compassionate care of the women who were running the Souriyat Across Borders center in Amman, a hospice for those wounded in Syria. The warmth and love exchanged were heartwarming to witness.

The warm and nurturing environment offered Habib opportunities to adapt to the major challenges of his catastrophic injuries and trauma. With unwavering empathetic support of the rehabilitation center volunteers, his healthy development was made possible through the loving family- and village-like atmosphere. He was carried around the center with the same affection a highly nurturing parent would offer a baby. Though we were visitors, not core staff, his responsiveness to interactions with my colleagues and I was heartening. He apparently was developing an ongoing, coherent sense of self that allowed him to establish new relationships based on trust.

We can only speculate that he had previously formed a secure attachment to his parents before they were killed. This may have promoted an infrastructure of psychological resources that enabled him to respond to the warmth and care provided at the center. Habib had no explicit memory of the traumatic incidents. Based on the description of his anxiety and fear when he first arrived, he had made dramatic changes and was no longer hypervigilant and agitated. He was clearly building trust and was thriving in that loving environment.

The degree to which his implicit memory of the complex trauma dominated less was reflected by the way he engaged in the world. He appeared to possess a flexible autonomic nervous system through which he could turn on his sympathetic system when needed and his parasympathetic after the challenge have been met. Presumably, the regulation of the various calming neurotransmitters systems such as the endorphin and benzodiazepine receptors allowed him to recoup when needed and self-soothe. Flexible cortisol regulation permitted him to ramp up the neurochemistry to deal with a challenge during the daytime, and then during the evening he could turn down cortisol so that he was better able to restore resources, enhance positive immunological functioning, and promote better neural growth and plasticity for cognition.

> One of the principle goals of therapy with children, adolescents, and adults is to promote secure attachment, to bolster not only interpersonal skills but also affect regulation and cognitive skills.

For Benny, who had developed significant mistrust of others, building trust was fundamental. The new learning was often punctuated by extinction bursts so that he often felt more mistrustful before feeling at ease with his foster parents and me. His defensive posturing, avoidance, and pushing me away before anticipating being abandoned again slowly dissipated until it became clear that it no longer worked and was not necessary in the new, safe context. Therapy oscillated between trust and no trust until trusting resulted in no bad outcomes.

In contrast, Habib had a therapeutic community embracing him. Despite his horrific trauma, both psychologically and physically, he received robust support through constant loving attention. Benny's loose-knit foster home and weekly therapy did not compare. I had to weave together and promote a semblance for Benny of what Habib had. From a Western "clinical" perspective Benny was receiving "appropriate treatment," while Habib was not, but it was dramatically lacking.

VARIATIONS IN ADULT ATTACHMENT

With all the attention focused on childhood, do residual effects of attachment extend into adulthood? Indeed, it appears that secure attachment noted in childhood is associated with a lower incidence of psychological disorders in adulthood, while on the other extreme, disorganized attachment is associated with dissociative problems. Anxious/ambivalent and avoidant attachment styles are both associated with the development of depression during adulthood. While avoidant style promotes depression based on a sense of alienation, anxious/ambivalent style promotes depression based on an internalized sense of helplessness and doubt (van IJzendoorn & Bakermans-Kranenburg, 1996).

In an attempt to assess adult attachment styles, Mary Main and her colleagues have developed the Adult Attachment Interview (AAI), based on eighteen questions assessing adult attachment history. There appears to be a correlation between attachment styles in childhood and adulthood. For example, the AAI-identified preoccupied style shows strong similarities to the ambivalent style. Ambivalent babies tend to have preoccupied parents, and they later become preoccupied adults themselves who are obsessed with love and loss. They tend to be emotionally underregulated and prone to overreact to interpersonal stress. Attachment predicts adult capacity for intimacy and conflict resolution. A meta-analysis of the AAI studies indicates that insecure attachment is correlated with anxiety and mood disorders (Cassidy, Jones, & Shaver, 2013).

Children (Infant Strange Situation)	**Adults (Adult Attachment Inventory)**
Secure	Free/Autonomous
Avoidant	Dismissing
Ambivalent	Preoccupied
Disorganized	Unresolved

The preoccupied pattern typified the predicament for Carl. He repeatedly checked with his partner, Rob, about the look on Rob's face that Carl interpreted to mean that "he was done me." Though Rob immediately reassured him that all was okay between them, Carl would follow up by asking, "You're sure about that, right?" Frustrated with the persistent questions, Rob sighed deeply. Carl threw up his hands and said, "You see! Why are you sighing? What did I do?" Rob told him to "get help for the sake of our relationship," which brought him into therapy.

The AAI-identified dismissive style shows strong correspondence to the identified avoidant style of attachment identified in children in the Infant Strange Situation. Avoidant children grow up to be dismissive adults, learning to deal with parental insensitivity by shifting attention away from their emotions. They tend to be little aware of the feelings of others. Dismissive adults tend to state that they remember little of their childhood, not because of repressing difficult memory but because few emotionally relevant events occurred while growing up. Dismissing adults tend to be overregulated and rigid even though separations still provoke spikes in cortisol.

The dismissing attachment dilemma was illustrated by Sharon and Tyler in a couples counseling session. Sharon complained that Tyler was "not there for me." He snapped back, "I'm with you all the time, except when I'm at work! What do you want me to do, quit work?"

"No, I mean that you are remote."

Tyler rolled his eyes and then turned to the couples counselor for an explanation. The counselor asked him what his parents' marriage had been like. Tyler shrugged his shoulders. "Fine, I guess. No abuse, if that's what you're asking."

Still not satisfied, the counselor pressed further. "Can you tell me more details of your childhood and family life?"

Tyler shook his head: "I really don't remember much of my childhood."

"Birthdays, holidays?"

"Just like any other days." His parents reportedly did not make those days special in any way—no parties, no family gatherings. With very few emotionally relevant events, every day looked the same, with nothing out of the ordinary to remember.

Avoidant as a child and dismissive as an adult, Tyler saw his physical presence with Sharon as meaning he was there all the time with her—it was that simple for him. But not for Sharon. Tyler possessed very little self-awareness, so he had little awareness of Sharon's emotional needs. His salience network was underdeveloped. Learning to be aware of his own emotional feelings facilitated sensitivity to Sharon's.

ATTACHMENT AND THE MIND'S OPERATING NETWORKS

As noted in Chapter 1, the organization of a coherent and durable sense of self is maintained by the mental operating networks. Clara's dysfunctional adult relationships undermined the development of and balance among those networks. When a relationship became emotionally stressful, she pulled her away, trying to avoid the pain of what felt like rejection. Her executive network was compromised by attention turned away from the present moment and instead inward for self-reflection with her default-mode network (DMN), which she used to replay the story lines of conflicts and to anticipate more of them.

Because Clara developed a dismissive attachment style, her capacity for insight about others was compromised. She found reflection about relationships difficult, in part because of underdevelopment of both her DMN and salience network. If she had been traumatized, switching between her compromised executive network and DMN may result in impairments in the functional connectivity between them (Daniels et al., 2014). Overall, therapy promotes flexible and socially adaptive shifting between her executive network, salience network, and DMN.

Increased coupling of the salience network and DMN has been associated with egalitarian, self-sacrificing, and prosocial behavior (Dawes et al., 2012). The DMN does not begin to develop until the early latency period, to participate in the emergence of social skills. It also offers a mental platform to link social cognition and self-referential thought (Mitchell, Banaji, & Macrae, 2005). As the network that developing children cultivate thoughts about themselves while reflecting upon their interactions with others, Clara used her DMN to maintain continuity while increasing complexity in her relationships: our DMN provides a means to reflect on our relationships, and our salience network provides a means to empathize.

These mental networks toggle back and forth to meet social demands. A burst of activity in the salience network signals important changes in the body in response to social situation and that it is time to snap out of the DMN and activate the executive network to make whatever adaptive behaviors are appropriate. A change in the social environment that may result in the recognition in social inequality and rejection signals the salience network to work with the DMN to reflect on what occurred and the feelings associated with it (Masten, Morelli, & Eisenberger, 2011).

A secure adult flexibly shifts from the DMN to the executive network to focus attention on cognitively demanding tasks while maintaining activity in the salience network for self-awareness. In contrast, people who are not able to shift flexibly among networks sometimes have trouble focusing on and adapting to changing circumstances in their immediate environment. This is what happened to Scott, as I discovered during his evaluation for attention deficit disorder. Interestingly, he held a graduate degree in chemistry, which he earned with high honors, and he had no difficulty with attention when studying alone. But he did find it difficult to focus on studying when people were in the room, even if they were very quiet.

Scott's developmental history revealed much of the context to his problem with focus. He was the last born of a family of eight children and was told that he was an "accident." Because his parents made it clear that they were "done" parenting, his siblings became his surrogate parents. Yet, they were also children and engaged in their own sibling rivalries. Scott grew up hypervigilant to all of their conflicts to ensure that he would not get roped in. As an adult with a preoccupied style of attachment, his external locus of control in the presence of others compromised the balance and flexibility between his executive and salience network.

RECIPROCAL SELF-AWARENESS

Implicit memory forms the undercurrent of interactions among family members who can share a brief glance that means much more to them than to observers outside the family—these are learned by children as part of development. Implicit communication patterns resonate to increasing degrees in people who have spent the most amount of time with the family. When they form friendships, get married, and have close work relationships they can run the risk of overinterpreting or underreacting to implicit nonverbal interactions with others outside their social networks. It is the task of therapy to discover these patterns and help clients use and modify them to conform to flexible healthy relationships.

> A long-standing practice in therapy is to make note of a client's expression, a sigh, or crossing of the arms that might denote a sudden implicit-emotional shift in the relationship. A therapist may say, "Look, what just happened between us?" This question punctuates the moment, making it memorable, and shifts the focus to the implicit attachment dynamics inherent to the therapeutic relationship.

The earliest recollections of struggles for love and safety may not be apparent in explicit memory but can be accessed implicitly through the transference relationship (Cozolino, 2017). The capacity to maintain self-awareness while reflecting on the thoughts, emotions, and behavior of others represents a key goal of therapy. These skills are dependent on our ability to consider another person's point of view while reflecting on our own.

The overlapping concepts of interpersonal relatedness:

- Theory of mind
- Mindsight
- Intersubjectivity
- Mentalization
- Attunement
- Social intuition
- Social intelligence

According to outcome research, clients tend not to completely reveal their feelings about therapy when asked in the session. Follow-up outcome questionnaires have become standard practice to discover negative concerns that the client has about the therapy that have not been addressed. With this new information, repairing the impingements and working through and earning security can be addressed.

The implicit feeling of therapy has a major effect on what clients reflect on between sessions. Derived from their implicit memory system representing their emotional tone, clients build nonconscious knowledge of relationships that tend to be continually replicated. Gaining insight into how these emotional patterns are replayed helps change the narrative in the explicit memory system. Implicit patterns of relating are learned slowly through healthy, (good-enough) relationships. Heightened affective moments signify the coregulation of shared implicit meanings (Beebe & Lachmann, 2002).

STORYTELLING AND THERAPEUTIC NARRATIONS

Since the advent of language, telling stories has served multiple functions. As a means of enculturation, members of a society can agree on a common origin and meaning to their existence and establish and reinforce morals, ethics, and general customs for societal cohesion. Storytelling, sometimes called the narrative, provides a means through which a therapist helps a client to organize adaptive self-referential meaning.

Stories serve a fundamental role in the interaction between parent and child. Storytelling represents an important part of the bedtime ritual. Stories capture attention if laced with suspense, drama, and tension and can be the most comforting way to learn life lessons when offered with resolution. They carry an emotional arc fused with cortisol in the beginning; the tension rises, with the child's attention riveted, eyes widened, heartbeat quickened, and then a resolution is facilitated by a brave and kind hero, with the child enjoying the climax with boosts in dopamine and oxytocin. This ritual offered on a regular basis primes the oxytocin system, increasing trust with the storyteller (Zak, 2012). Because listening to stories brings the combined release of cortisol and oxytocin when the plot factors in empathy, it puts limits on the elevation of cortisol. Like therapy, the story teaches how social support and resiliency can work together.

Children and their caregivers co-construct narratives that describe and helps children make sense of their existence. Through co-constructing narratives children learn how to develop meaning and a sense of belonging. These narratives contribute to the formation of the explicit memory system within which autobiographical and episodic memories contribute the self-referential information, which the DMN taps to reflect on, to develop meaning in relationships.

Therapy entails co-constructing adaptive new narratives to be replayed during clients' DMN periods. These storylines instill social support, slowly building trust in people who deserve trust, and a sense of security. Through the self-referencing narrative, clients learn to imagine conversations with socially supportive storylines. Co-constructive narratives offering uplifting emotional arcs, soothing storylines, or Zen koan-like stories entice clients to search creative solutions to their problems. Having heard a story seemingly related to someone else but in actuality meant for the client can be imbued with a solution focus (O'Hanlon & Rowan, 2003).

Self-referential story loops comprise narrations that we tell ourselves about our relationships. Distressing events are not dwelled on in the DMN for securely attached people as much as they tend to be with insecurely attached people. Sad stories tend to have a happy or at least hopeful ending for the securely attached, whereas an insecurely attached person fixates on negative aspects of a relationship.

> Creating unexpected moments of empathy and compassion can trigger positive prediction errors. When a client sees you smile unexpectedly, it induces a release of dopamine, which is received by her nucleus accumbens so that she becomes motivated to create the conditions for that to recur. Such moments create a window of opportunity to coconstruct new narratives that build trust and hope.

In family therapy parents can learn to develop warm and engaging expressions toward their children, while storytelling fosters trust (Baylin & Hughes, 2016). The stories may include information about family conflict needing to be resolved. When the parents stay open and engaged with each other and with their children, the tension or misattunement can be worked through.

Narratives co-constructed in therapy are best if multilayered, because clients experience innumerable events in their life, and the contexts are constantly changing. No single narrative modeling their experiences will suffice. Since they interact with multiple people, flexible and adaptable narratives are effective. The fluid and changeable nature of their social experiences requires narratives that are durable with multilayered social contexts.

Collectively, the social factors explored in this chapter illustrate why the quality of the therapeutic relationship has been consistently found to be the most significant factor affecting outcome. For this reason, some have borrowed the real estate cliché to say that the three most important factors in therapy are relationship, relationship, relationship.

CULTURAL VARIATIONS

We coevolve with others within an interpersonal field, community, and culture that all provide the social feedback loops that offer context and meaning to our experience. Based on these culturally bound contextual reference points, we communicate to others and set up expectations about the values or the potential effect of our behavior. From infancy on these culturally relevant factors frame our relationships.

Infants born to parents in northern European cultures respond to adults who are not their parents differently than do infants born within an Efe (Pygmy) family. While infants in northern Europe may find it difficult to be comforted by anyone other than their parents, an Efe infant probably would find it easier to interact with everyone on a daily basis. These individuals have multiple attachments, so there are sharp distinctions between self and other (Tronick, 2007).

The Infant Strange Situation paradigm has been used throughout the world, revealing a preponderance of particular ethnic attachment styles.

For example, in northern Germany and Scandinavia there are relatively more people with avoidant patterns of attachment. It is not uncommon for mothers to step away briefly, leaving their infants unattended at home or outside of supermarkets. A well-publicized incident typifies this generalization when a Scandinavian couple visited Manhattan. They left their baby in a stroller outside the supermarket and went in to purchase food. Child Protective Services arrived like a SWAT team, applying a protective perimeter around the baby. The parents emerged from the market and were alarmed to find people hovering around their baby. When the Child Protective Services workers asked why they left their baby outside, they responded with astonishment, saying "What do you mean? We do that at home all the time!"

In Japan there is a preponderance of ambivalent and hard-to-soothe infants. Mothers and infants are rarely separated. Babysitting is rare, and when it occurs it is generally with grandparents (Miyake et al., 1985). Among kibbutzim in Israel, babies have been reported to become anxious by the entry of strangers in attachment testing situations. Terrorist attacks create a xenophobia reflected in their parents and modeled by children, so that strangers are distrusted (Saarni et al., 1998).

When I lived on a kibbutz in the early 1970s I was intrigued by the group child-rearing practices. Instead of children living with their parents, they lived in separate quarters, like a dormitory, and the adults would take turns taking care of them. Bruno Bettelheim argued that this practice was meant to rid them of the inward-oriented, self-conscious personality of the ghettos of Europe and help them become more oriented toward modern society. With an extreme deemphasis on the roles of their parents, the children presumably developed multiple attachments. However, in retrospect, attachment was compromised.

Within the Gusii culture in Kenya, a mother does not often engage her infant in face-to-face play. When mothers are asked to engage her infant face to face, just as the infant responds excitedly, the mother turns away. In contrast, among the !Kung in the Kalahari infants spend 70–80 percent of their first year in constant physical contact with their mother. The remainder of the time they are handled by someone else.

They are nursed four or more times an hour for short bouts (Konner & Worthman, 1980). Babies and the adults engage in animated face-to-face interactions.

Given the wide range of diverse interpersonal styles of relating, the practice and training for psychotherapy have attempted to adjust accordingly. Approximately forty years ago, while teaching counseling on the Navajo Reservation, I adjusted the conventional expectations regarding relationship-building skills, such as eye contact. Among Navajos, eye contact was discouraged and suggests disrespect. Also, in contrast to the Rogerian and later motivational interviewing concept of active listening, the Navajo conversational style was relatively more authoritarian, with the counselor telling the client what they ought to do. During the period I wrote this chapter I gave a seminar in Albuquerque and a Navajo counselor who attended approached me during the break. He maintained engaging eye contact as he asked a question. I noted that his eye contact was strong, and I asked him about the traditional practice of having little eye contact. He responded with a broad smile and said, "I'm well acculturated in Anglo society."

Because of these cultural variations in communication styles, almost all psychotherapy training includes cultural competence as part of the curriculum. The training programs that I directed in twenty-four medical centers were all required to weave cultural competence into every aspect of training.

> As psychotherapists we should not only be sensitized to the culture but also be aware of the implicit biases and microaggressions that we may make with clients of diverse backgrounds, whether it be culture, race, or LGBTQ differences.

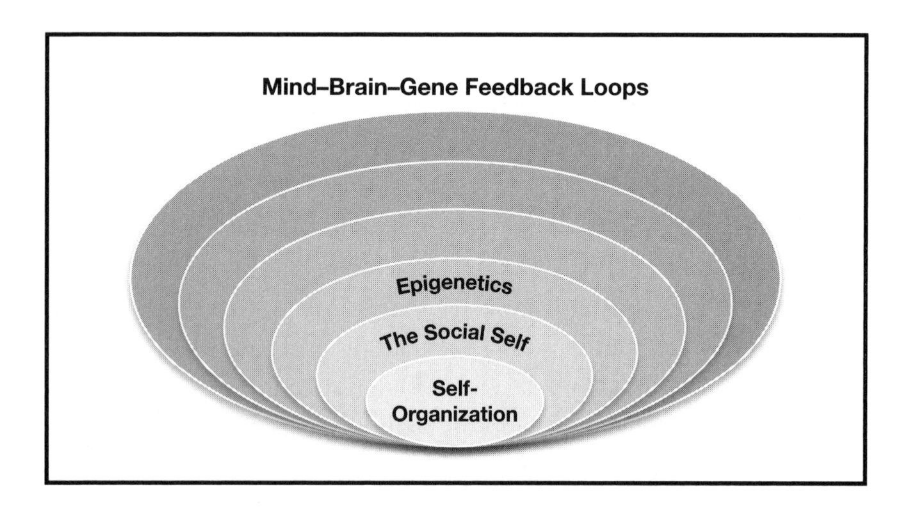
Mind–Brain–Gene Feedback Loops
Epigenetics
The Social Self
Self-
Organization

CHAPTER 3

Behavior-Gene Interactions

> Psychotherapy works by producing changes in gene expression that alter the strength of synaptic connections.
>
> —Eric Kandel

To provide a truly integrated psychotherapy we must expand our scope to include relevant groundbreaking new developments in science. One such revolutionary development includes the new field of epigenetics. Recent studies in epigenetics have shown that the quality of care early in life has a significant effect on gene expression. Consistent with the attachment research explored in Chapter 2, this chapter explores how good early nurturance can increase stress tolerance and how poor care undermines it through epigenetic changes, as well as how acute and/or chronic stress during childhood can result in not only psychological disorders during adulthood but also major health problems through epigenetic effects.

Yet, there is hope. Psychotherapy can influence how genes interact with the environment, turning on or off by experiences and behavior. As Eric Kandel pointed out in the epigraph to this chapter, psychotherapy produces changes in gene expression that change how the brain functions. Genes are expressed or suppressed within the context in which we live, and our development, self-care, behavior, and decisions shape that context.

We have come a long way from the brainless, purely theory-based psychotherapies of the twentieth century to now understand how promoting behavior change results in either gene expression or suppression. How did we get here? And more important, how can we convince clients who feel hopeless, who feel they are doomed to misery because of their genes, that there is hope?

The way we think about the role of genes needs to change. Gina, for example, had always believed that she had been dealt "bad genes." Born of teenage parents who immediately gave her and her identical twin away to adoption, she spent much of her childhood feeling rejected by both her biological and adoptive parents. She never remembered being held, kissed, or told by her adoptive parents that they loved her.

As Bosnian refugees who had witnessed and experienced horrific violence, both adoptive parents were often too distracted by the past to be completely present for her. She was emotionally neglected, at great cost to her health. In fact, on several occasions she overheard them mutter, "After what we've been through adopting her was a big mistake!" They told her that she was a colicky baby and that "every day one of us drew the short end of the straw to decide who had to hold you." The truth was, her father never did, and her mother typically held her for a moment before plopping her back into the crib. Her father blamed his alcoholism on her and punished her mother with verbal abuse and physical violence for pushing the idea of adoption. Her mother became increasingly depressed and despondent. Gina remembered her as an "empty shell." The only attention she received from her father was emotional torment. During his drunken rages he repeated that "your mother is depressed because of you!"

Throughout adolescence and into early adulthood Gina suffered from a range of health problems, including obesity, type 2 diabetes, anxiety, depression, and fibromyalgia. Her mental operating networks were significantly imbalanced, with her salience network generating what she described as "icky feelings." She said that "comfort food" provided her a "go-to" method for feeling relaxed. She would retreat to her room with excessive default-mode network activity, where she sat ruminating that her biological parents must have passed on "bad genes" to her and this was the reason to give her up for adoption. Her executive network activity was focused primarily on her parents' moods and not on developing goals for herself. Through the support of her therapist she tried to track down her biological family to "discover what illnesses and psychological traits were genetically passed down." She was only able to find her identical twin, Sara, who did not have any of these problems. She was adopted by parents who offered a loving and stimulating family environment. If genes were playing a role in Gina's poor mental and physical health, they did not appear to result from what she received from her biological parents. But genes may have been affected by the neglect and emotional abuse of Gina's adoptive parents and her own behavior. Her stress tolerance, mood, and self-care were all interconnected with gene expression. This chapter explains how.

ADVERSE CHILDHOOD EXPERIENCES AND THEIR HEALTH CONSEQUENCES

The history of psychotherapy is peppered with somatic-based theories and anecdotal reports of people like Gina who experience trauma and then later become physically ill. Beginning with Jean Martin Charcot's idea of female hysteria, followed by Freud's sexual repression theory, and then conversion disorder and somatic medicine, the mind-body connection had yet to evolve into a comprehensive model and beyond broad generalities. We are now better able to understand and change the interactions among the mind, the brain, and genes. By understanding the pervasive impact of psychological experience on gene expression, physical health and mental health can now integrate health care.

Large and comprehensive health surveys have demonstrated a profound link between early psychological adversity and physical health problems. For example, Vincent Felitti of the Kaiser Permanente Medical Centers, in collaboration with the Centers for Disease Control and Prevention, conducted a landmark study involving over 17,000 middle-class adults to explore the effects of childhood adversity on adult health. Now known internationally as the Adverse Childhood Experiences (ACE) study, it examined the relationship between stressful and traumatic experiences and their impact on long-term health.

The ACE study illustrated how ten categories of adverse experiences in childhood have destructive impacts decades later on a wide variety of health risks, social problems, and reduced life expectancy:. The ACE categories include growing up in the family with various types of abuse, including incest, alcoholism, drug abuse, mental illness, criminal behavior, and/or domestic violence. ACE scores were tabulated by summing up the number of categories experienced, not the number of incidents or events within one category of an ACE. Less than half of all the participants, like Gina's twin, Sara, had an ACE score of zero; like Gina, one in six had an ACE score of 4 or more. ACEs are more common than not, with two-thirds having an ACE score of at least 1.

Category	Prevalence
Childhood Abuse	
Psychological (by parents)	11%
Physical (by parents)	11%
Sexual (anyone)	22%
Household Dysfunction	
Substance abuse in family	26%
Mental illness in family	19%
Domestic violence	13%
Imprisoned household member	3%
Loss of parent	23%
(Centers for Disease Control and Prevention https://www.cdc.gov/violenceprevention/acestudy/ace_brfss.html)	

The study matched the ACE score with health risks such as heart disease, diabetes, obesity, depression, and suicide. With each health category the risk increased for a stepwise increase in ACE score, with no exceptions. In other words, the more ACEs a person experienced, the greater the likelihood of health problems developing in adulthood. Mental health professionals have long believed that childhood adversity could result in psychological problems, including anxiety and depression. So it is not surprising that increases in ACE scores were associated with serious social and psychological problems. It is the degree of impairment that is compelling. For an ACE score of 3 or more there are increases in marital problems, and for an ACE score of 5 or more, increases in sexual problems. For people with an ACE score of 7 or higher, compared to people with an ACE score of 0, there is a 3,100 percent increase in the incidence of attempted suicide! Overall, the ACE study demonstrated that social and health well-being can be profoundly impaired as much as fifty years later in the form of physical illness, social deficits, and psychological disorders.

ACE Correlations With Adverse Outcomes

As the ACE score increases, so does the risk for the following:

- Obesity
- Diabetes
- Coronary heart disease
- Myocardial infarction
- Stroke
- Asthma
- Anxiety
- Depression
- Mental distress
- Smoking
- Disability
- Lower reported income
- Unemployment
- Lower educational attainment

(Centers for Disease Control and Prevention https://www.cdc.gov/violenceprevention/acestudy/ace_brfss.html)

Poor self-care and destructive habits are associated with ACEs and add to the breakdown of health. With an increase in the number of

ACEs, the risk increases for engaging in destructive relationships, substance abuse, smoking, alcoholism, and drug use. For example, for alcoholism we see a 500 percent increase for people with an ACE score of 5 or greater, compared to those with a score of zero, and with a score of 6 there is a 4,600 percent increase in the likelihood of becoming an intravenous drug user.

The interactions between childhood psychological stress and adult health problems contribute to significant economic costs to society. The overall public burden resulting from ACEs has so garnered the attention as of 2009 that policy makers in eighteen US states are routinely gathering ACE information. Internationally, the Centers for Disease Control and Prevention and the World Health Organization are testing the usefulness of ACE questions in health surveys.

THE CALL FOR BEHAVIORAL HEALTH

In a prod to his colleagues in medicine, Felitti has pointed out that the ACE study demonstrates the overall importance of an integrated approach to health care that includes a focus on mental health: "Our findings also make it clear that studying any one category of adverse experience, be it domestic violence, childhood sexual abuse, or other forms of family dysfunction is a conceptual error. They do not occur in vacuo; they are part of a complex systems failure: one does not grow up with an alcoholic where everything else in the household is fine." This call for integration is critically relevant to mental health professionals, urging us to embrace general health care and go well beyond what was traditionally the limits of training in psychotherapy.

Because there are complex nonlinear relationships among the biopsychosocial factors impacting each person, the psychotherapy of the twenty-first century must contribute to decompartmentalized health care. It behooves all health care providers to understand the interrelationships among ACEs, psychological challenges later in life, and a wide variety of health problems. For example, 10–15 percent of cases in the ACE study had both a high ACE score and coronary heart disease yet had none of the risk factors for the disease, according to the gold-standard Framingham Heart Study. Individuals with ACE scores of 6 or more tend to have shortened life expectancy by almost twenty years.

Trauma survivors are significantly more likely to have a wide variety of serious illnesses and die prematurely than are nontraumatized people (Felitti et al., 2001). The well-known Dunedin Multidisciplinary Health and Development Study, a longitudinal study in New Zealand, followed people who had experienced childhood maltreatment and found that they had elevated levels of inflammation twenty years later. The severity of the abuse is positively correlated with the level of inflammation. Based on data derived from almost 37,000 subjects participating in the Canadian Community Health Survey, those suffering from posttraumatic stress disorder (PTSD) had significantly higher rates of cardiovascular disease, chronic pain syndromes, gastrointestinal disorders, respiratory diseases, chronic fatigue syndrome, and cancer (Sareen et al., 2007). Consistent with these findings, the ACE study analysis of twenty-one different autoimmune diseases revealed a significant proportionate relationship between ACE score and the likelihood of developing autoimmune diseases later in adult life. (Interactions among the mind, brain, and immune system are addressed in Chapter 4.) Thus, illnesses that had in the past been considered the result of genetic destiny are now understood as having bidirectional causal relationships with psychological factors.

How do psychological and social factors interact with genes of genetically identical people such as Gina and Sara, and produce a wide range of different health effects? How was it that Gina became both easily stressed and less able to anticipate pleasure and to self-sooth? How did she find relief through eating comfort foods, only to narrow her options to soothe herself in more adaptive ways? Sara, by contrast, did not feel the need to comfort herself in those ways. She was resilient, in part, because genes that were expressed in Sara were the same genes suppressed by Gina. To examine how genes interact with the environment and our behavior, it is important for psychotherapists to understand how and why our understanding of these effects has changed.

THE DEMISE OF GENETIC ASTROLOGY

The idea that ACEs could produce a plethora of illnesses did not fit into the zeitgeist of mainstream health care until recently. Neither did the idea that health problems contribute to mental health problems. For that matter, the idea that life experience could change how genes turn on or off sounded like mere fiction.

Mental health services had been dominated by the reductionist medical model that envisioned psychological disorders as caused by possessing specific genes, which are associated with imbalanced neurotransmitters that need to be "treated" with the "right" medication. Therapists were pushed to the back of the bus in favor of medical professionals who provided medication designed to "rebalance" neurotransmitters to correct for "bad genes." Gina and her therapist bought into this mindset, only discover that things are not that simple. Genetics had been assumed to be the primary causal factor in health, so this section starts at this level and describes how we have come to understand that genes can be turned on or off.

By the turn of the twenty-first century researchers were pointing to the gross limitations of genetic reductionism. The simplistic idea that we have a gene for this and a gene for that has been referred to as "genetic astrology" (Jablonka & Lamb, 2006). When considering psychological disorders and psychotherapy, genetic reductionism makes little scientific sense. We do not have specific genes to account for all the psychological disorders catalogued in the latest version of the *Diagnostic and Statistical Manual of Mental Disorders*. And if we did possess specific genes for specific disorders, psychotherapy would have limited benefit.

Taking genetic reductionism to an extreme would mean that we should be much more similar to chimps than we actually are in behavior and consciousness—after all, we share 98.7 percent of the same genes. But our behavior is not 98.7 percent identical. Even the number of genes that we possess makes the idea of genetic reductionism seem archaic. It was believed as recently as in the 1990s that there were as many as 100,000 protein-coding genes in our DNA. Thanks to the Human Genome Project we now know that there are only 20,000 to 25,000 genes in the human genome. *Caenorhabditis elegans*, a roundworm often used in genetic research, has roughly 20,000 genes, and obviously our behavior is much more complicated than worms. They do not have psychological disorders for which they seek professional help, nor do they read or write books about the complexities of their health problems. Clearly, we are much less determined by genotypes than such primitive creatures. In fact, for humans most genes are expressed in only a minority of tissues, whereas upward to 60 percent of our genes are expressed in the brain. This means our brain has influence over a larger percentage of genes than do other body organs, allowing significant potential to shape multiple levels of mind-brain-gene interactions.

It turns out that genetic effects are more complex than once assumed. Genes are generally defined as sequences of DNA that code for protein. Yet, only 2% of our DNA codes for protein, and 98% were mistakenly called "junk DNA." Now we know that these portions of DNA are not junk at all. The greater the amount of junk DNA, the more complex the species. A worm's behavior is obviously far more driven genes than ours.

In fact, there are a wide variety of functions performed by "junk DNA." For example, genes can be switched on through "junk DNA" referred to as promoters and enhancers. Within the immune system, once an inflammatory stimulus has been removed, enhancers do not revert back (turn off) to their original state. They continue as enhancers, ready to up-regulate an immune defense, should the threat return.

Despite these unanticipated complexities, genetic reductionism is still embraced by many in the general public and even many health care professionals. For Gina this mistaken belief brought with it a sense of hopelessness that she could not do anything about her situation. Her psychotherapist could have promoted hope instead of hopelessness by discarding the simplistic reductionism of the past and cultivating awareness of the current science.

To understand mind–brain–gene feedback loops, evolution is our guide. The bedrock of evolutionary theory involving genes has evolved. The traditional model of genetics had previously provided support for the gradualist perspective of evolution by natural selection. Those species evolving within changing ecological niches were "selected" by the environment because they survived to pass on their genes. It was thought that genes change not as a result of behavior but, rather, gradually as individuals that are most fit for their environment survive and pass on their genes. The popular expression of this concept is "the survival of the fittest." Richard Dawkins took genetic reductionism to the extreme with the publication of his book *The Selfish Gene*. For many this was more than a mere metaphor, meaning that everything was determined by what genes "want."

In fact, the tide shifted about the role of genes when evolutionary biologists pointed out that species change more abruptly than previously believed. The interaction between the behavior of animals and their environment produces evolutionary change. Actually, this idea was not new. Well before Darwin there were theorists who proposed that behavior played a significant role in evolution.

Jean-Baptiste Lamarck preceded Darwin with an evolutionary theory by forty years. Now referred to as the inheritance of acquired characteristics, the theory that Lamarck proposed was that the behavior of one generation served as the driver for evolutionary changes for the next generation. The theory was popularized by the image of a mother giraffe stretching her neck to eat leaves in a high tree and then later giving birth to offspring with longer necks. Lamarck's theory never gained support in the scientific community. Instead, Darwin's theory of natural selection became the standard model for evolution. It favored a gradualist perspective envisioning evolutionary change as occurring at glacial speed, over hundreds of generations. Eventually the theory was buttressed by the developing field of genetics. However, despite that the theory of natural selection conflicted with Lammark's Theory of Inheritance of Acquired Characteristics, Darwin was nevertheless quite impressed. He said, "Lamark . . . had few clear facts, but so bold and so many such profound judgment that he foreseeing consequence was endowed with what may be called the prophetic spirit of science. The highest endowment of lofty genius." Darwin believed that in addition to natural selection, the Inheritance of Acquired Characteristics was one of the mechanisms of evolution.

It turns out that although Lamarck's theory is a simplistic explanation of evolutionary change, the role of behavior has now found a new form of validation. Genes, the environment, and the behavior of the individual are interdependent and all coevolve—not quite giraffe mothers stretching their necks and giving birth to baby giraffes with longer necks, but gene modification can appear much faster than we once thought.

The reductionist medical model that promoted primacy of imbalanced neurotransmitters and bad genes held sway until the science of genetics was itself transformed by the discovery that genes can be turned on or off by the environment and behavior. To appreciate how psychotherapy plays a role in this sea change, we need only appreciate that genes do not exist in a vacuum. Genes express themselves within the context in which they exist; cells, the body, the environment in which the body lives, the behavior and health of the person whose body it is, and the decisions and emotions of that person.

Epigenetics: Above the Genome

Genetic change can occur within one lifetime or lead to change in the very next generation. One of the most noted examples of rapid genetic change occurred to the people born in Holland just after the infamous Dutch Hunger Winter. From November 1944 to late spring 1945 the Nazi regime punished the Dutch by blockading food aid during a particularly cold winter, causing many thousands to die of starvation. Women who were pregnant at the time gave birth to children who developed significant health problems decades after that winter. The key point is that the genetic effects of the famine were not immediately evident; they showed up when the offspring became adults. If the mother suffered from malnutrition for the first three months of pregnancy but then was well fed, she was likely to give birth to a baby with normal weight. However, decades later, that person would be twice as likely to develop major health problems, such as type 2 diabetes, high blood pressure, cardiovascular disease, and obesity (Stein et al., 1995). The body of the adult who had been starved in utero became genetically reprogrammed to abnormally grow extra fat cells to store energy for later use. Though the famine occurred only during the time they were in utero, as adults they were "prepared" for a potential famine by excess weight.

Though nutritional deficits, and certainly famine, have long been shown to cause cognitive deficits during childhood, it is the long-term modification of genes that illustrates a sort of Lamarkian evolution. In just one generation the children adapted to the famine conditions of the mother despite not encountering those conditions themselves later in life. These adaptations came at a significant cost, with an increased risk for mental health disorders, such as depression (Brown et al., 2000; Hoch, 1998). And what is most illustrative of rapid genetic change is that they went on to have children with similar problems.

The emergence of the field of epigenetics has revolutionized biological science, health care, and psychotherapy. Epigenetics, which means "above the genome," explores how and why some genes are expressed while others are suppressed. Our behavior and environment play critical roles in changing our biology even at this microlevel. Epigenesis represents an experience-dependent change to the expression of our genes. Because psychotherapy promotes behavior changes, it plays an important role in which genes are expressed (turned on) and which are not. When our experience changes in response to challenges, some genes are expressed to produce proteins that play a crucial role in the creation of new neural pathways to facilitate new adaptive skills.

How this occurs is no longer a mystery. Genes and the environment interact, with multiple layers of interdependent complex systems in between, including our mind, brain, immune systems, and cells. In fact, even cells that have the same set of genes can vary. They respond to internal and external conditions, turning genes on or off as they interact with their microenvironment. Cells change in response to our behavior and the environment we choose to inhabit. Sometimes our choices promote psychological disorders, but these are not reducible to genes.

Changes in gene expression occur when two genetically identical people, such as Gina and Sara, have nonidentical experiences, behaviors, health, and biological development. Each of them produced different epigenetic effects as their brains adapted to differing interpersonal, environmental, and self-care practices. All of these variables are involved in nonlinear interacting layers of feedback loops influencing which genes are expressed or suppressed. Environmental factors impact both the biological and the psychological feedback loops.

To get a better idea of how the field of epigenetics has revolutionized health care and psychotherapy, it will be useful to touch on the high points of genetics and then work backward to appreciate the interdependence of mind, brain, and genes. From conception to death, cells in our body divide and copy our entire DNA: the two strands of the double helix separate and serve as a template for the creation of another strand to match it. Because of the pairing principle, the new double-stranded DNA molecules have exactly the same base sequence as the parent molecule (barring mutations). Each of our cells contains three billion base pairs of DNA, half inherited from our father, and half from our mother, to produce our unique sequence, referred to as our genome.

Genes are made of deoxyribonucleic acid (DNA), the fundamental information source or basic blueprint in our cells. DNA represents a code, a language, composed from the four nucleotide bases, adenine, cytosine, guanine, and thymine, represented by the first letter of each: A, C, G, and T. These four bases join up and are organized by the base-pairing principle, whereby A is always joined to T and C always to G. They are often described as teeth on a zipper, with two strips facing each other, linking up according to the base-pairing principle. These two strips, or strands of the zipper, are twisted around to form a spiral, famously described as a double helix, held together by chemical bonds between the base pairs. If there is an error in the sequence, such as A trying to link up with G, the whole shape of the DNA is thrown out of sequence, just like a faulty tooth on a zipper. Such a flaw may result in a genetic disorder, making the person vulnerable to various illnesses, physical deformities, or particular sensitivities.

Genes interact with the environment when necessary by making messenger ribonucleic acid (mRNA) to provide a code to make proteins, which serve as the basic building blocks of cells. Some proteins take the form of enzymes and cytokines in the immune system. The environment and our reaction to it can express or suppress genes through a process known as transduction. Signaling molecules outside cells interact within cellular receptors to initiate a cascade of biochemical reactions inside cells, which stimulate protein "transcription factors" to activate gene expression. Transcription factors flag a coding region of a gene (a particular stretch of DNA) for transcription into RNA. The transduction of socioenvironmental influences into genomic responses is mediated by our mind-brain's perception of social conditions and the subsequent regulation of neurotransmitters, hormones, and immune system molecules that disseminate throughout our body to activate cellular receptors and transcription factors. If we perceive those conditions as threatening, gene expression can occur through our sympathetic nervous system's fight-or-flight and/or the neuroendocrine pathways.

The coiled strands of DNA wrap around a core of small proteins called histones, which are folded into compact ball-like shapes with floppy tails of amino acids. When researchers discovered this entire DNA-protein complex under a microscope, it appeared purplish brown, so they named it a chromosome, from the Greek *khroma*, meaning color. The stuff of chromosomes, including DNA, histones, and RNA called chromatin, is dynamic, not static, and is not limited to only one possible outcome. Changes to the chromatin can produce gene expression or gene suppression. One mode of gene expression, called acetylation, involves the addition of a chemical group called acetyl that acts to "unwrap" the histones to express the genes. In contrast, a mode of gene suppression, or "silencing," is called methylation because it involves adding a methyl group to the cytosine base, which keeps the histones tightly wrapped so that the gene is not expressed.

Epigenetic effects can occur in many ways in response to the environment and our behavior. Though in general DNA methylation turns genes off and histone acetylation turns them on, these processes are not crude on/off switches but more like volume controls. In the brain, gene expression can occur in response to drugs or trauma, whereas in cells lining the gut the pattern of epigenetic modification will alter depending on the amounts of fatty acids produced by the bacteria in our intestines. As Chapter 4 describes, our gut bacteria can effect our mood and quality of cognition.

The dynamic interplay among genetic endowment, behavior, thought, emotion, and environment illustrates how the reductionist concept of anxiety and depression as "mental illnesses" took the mental health system to a theoretical dead end. Genes do not necessarily result in specific mental health disorders. On the other hand, the interaction of ACEs, poor environmental conditions, drugs, or a wide variety of bad habits may influence the expression of genes and thus our brains, emotions, and clarity of cognition.

Gina and Sara started with the same genes, but from birth they had dramatically different life experiences. The resulting epigenetic effects amounted to different patterns of gene expression and suppression, creating profound differences in their mental and physical health and thus in their quality of life.

No Blank Slate

New insights from the field of epigenetics have shown how particular genes combined with adverse experiences are associated with poor health. Despite John Locke's and the later behaviorists' claim that we are born with a "blank slate," we are not. Yes, we are molded by our experiences, but unless we have an identical twin we do not all start with the same genome. Some people are more vulnerable than others to developing a range of psychological disorders. And these variations are associated with epigenetic effects under challenging conditions. The key point is that vulnerability is not the same as causation, nor are the vulnerabilities related to specific psychological disorders.

A variety of genetic vulnerabilities and potentialities that can epigenetically affect mental health, depending on our behavior, thoughts, emotions, social situations, and the environment. The interactions between genetic vulnerability and ACEs have been illustrated by people with differing levels of the neurotransmitter-metabolizing enzyme monoamine oxidase A, which metabolizes dopamine (associated with motivation), norepinephrine (associated with alertness), and serotonin (associated with mood). Individuals in the Dunedin longitudinal study and a group of people in Richmond, Virginia, who had low levels of expression of genes encoding monoamine oxidase A and also suffered from childhood maltreatment had a higher incidence of conduct disorder, antisocial personality disorder, and/or impulse control problems, tended to engage in adult violent crime, and had deficits in cognition and attention (Caspi et al., 2002; Foley et al., 2004). The absence of direct reductionist links between the gene for this enzyme and a specific disorder had led some to call it "everything but the kitchen sink gene" (Heine, 2017).

Another gene reported to interact with ACEs and correlated with nonspecific but poor psychological outcomes is referred to as the short version of the serotonin transporter gene. It was incorrectly dubbed the "depression gene" in the mainstream press, but the truth is far more complicated. People with the short version may tend to exhibit greater amygdala activity (associated with threat detection) when exposed to fearful visual stimuli than those who have the long version (Hariri et al., 2002). Yet, culture plays a major role in the way this gene is expressed. The short version is possessed by 80 percent of Japanese, 59 percent of Indians, 45 percent of Americans, and 28 percent of South Africans (Chiao & Blizinsky, 2010). Because the short version is associated with lower levels of serotonin, if both Gina and Sara possessed this gene they both may tend to overrespond to emotional stimuli if they also had suffered from

ACEs, but only Gina had. If Sara also possessed the short version it did not result in developing anxiety disorders or depression.

> ***Epigenetics: For Better or Worse***
>
> Epigenetic factors play a plasticity role in enhancing the effects of nurturance or exacerbating the effects of ACEs. For example, infants with a particular variant of the dopamine receptor gene (DRD4) have been linked to lower receptor efficiency and greater risk for disorganization and externalizing behaviors if exposed to maternal loss or trauma. Yet, when children with this supposed "vulnerability gene" were raised by mothers who had no unresolved loss they displayed significantly less disorganization. With nurturing mothers, they show the lowest levels of externalizing problem behavior (Bakermans-Kranenburg & vanIJzendoom, 2007). In other words, having this variant of the DRD4 gene can afford the carrier to benefit disproportionally from supportive environments.
>
> The serotonin-transporter gene (5-HTTLRR) differentiates those people with the "short version" from the "long version" (eg S/S, L/S, or L/L). Since the short version was considered to be related to depression, either directly or in the face of adversity,the mainstream press dubbed it the "depression gene." While carriers of the short version have been associated with depression if they experienced ACEs, those who had experienced supportive early environment and positive experiences had the fewest symptoms (Taylor, et al., 2006).
>
> The genetic polymorphism BDNF alone does not operate as a plasticity factor, but the environment and multigene interactions together do. Thus, there are a variety of epigenetic factors at play influencing the differential susceptibility "for better or worse" to environmental influences (Belsky, 2013).

The point is that possessing the serotonin transporter gene may contribute to the vulnerability to, not the cause of, depression or generalized anxiety disorder if life experiences are particularly stressful, like Gina's (Kendler, 2005). Sara's secure attachment contributed to the development of good stress management skills and healthy lifestyle factors, which blunted her stress reactivity and minimized the risk of developing these disorders. Had Gina possessed the short version but had been adopted like Sara by parents who cultivated a secure attachment and taught her good stress management skills and positive lifestyle behaviors, she likely would have been more resilient. Stress can impact extracellular and intracellular signaling and gene-expression pathways in the brain. Gamma-aminobutyric acid (GABA) functions as the principal inhibiting neurotransmitter in the brain. It also serves as a major modulator of activity of the hippocampus (Reul, Collins, & Gutierrez-Mecinas, 2011). Because the hippocampus is critically involved in explicit memory processing, this interaction provides evidence for a molecular role of the emotion/cognition interface in stress-related learning and memory. Persistent stressful events may suppress gene functioning so that subsequent adult responses to stress are altered; these epigenetically turned-off genes might be transmitted to ensuing generations. Epigenetic processes affecting both cortisol and GABA receptor functioning are highly related to stress response, anxiety, and depression (Szyf, 2011).

Substance Abuse and Epigenetics

So far I have focused on epigenetic effects that occur in response to different early-life experiences, including ACEs. In addition to those factors, people may respond differently to various drugs, which can also produce epigenetic effects, especially when an individual possesses specific types of genes. Some adolescents develop psychotic symptoms after smoking cannabis, while others do not. Matt and Kyle both started smoking cannabis during high school, and their senior year, as their grades began dropping, so did their motivation to develop a plan for after graduation. Matt was having too much fun going to parties, having a wide variety of girlfriends, and spending hours at the beach. Kyle went to the beach with him but spent more time walking off alone, away from all their friends. He began feeling increasingly suspicious of their intentions, until those loose thoughts morphed into paranoid scenarios. Eventually he began to hear voices, which reinforced the paranoid thoughts. Had Kyle developed these symptoms forty years ago, the prevailing model assumed by many mental health professionals was that those psychotic symptoms reflected his "latent schizophrenia."

We now know that there are dynamic feedback loops between and within each dimension of the mind-brain-gene continuum that, depending on the combination of genes, behavior, and cannabis use, results in psychotic symptoms. For example, individuals in the Dunedin longitudinal study carrying a specific mutation of the *COMT* gene, like Kyle, who used cannabis during adolescence were more likely to develop psychotic symptoms in adulthood, while individuals like Matt, who did not carry this *COMT* gene variation, were less likely to develop psychosis (Caspi et al., 2005). In the brain, the catechol-O-methyltransferase (COMT) protein is of particular importance in regions such as the prefrontal cortex, which is typically dysregulated in schizophrenia. The point is that the *COMT* gene is not a "schizophrenia gene" but is an enzyme that breaks down dopamine, norepinephrine, and epinephrine. Kyle's psychosis emerged because his prefrontal cortex was not able to organize thoughts and feeling produced by the cannabis. Unlike Matt, Kyle was unable to frame cannabis thoughts and feelings as "fun."

Matt gradually lost interest in smoking cannabis as he shifted to alcohol as his drug of choice. He said that "partying" with alcohol made him more social. He began to consider drinking as a social lubricant. Eventually, he framed social situations without alcohol as not being fun. He also found that the more he drank, the more he needed to drink to get the same "sense of social comfort." This is because alcohol abuse, too, can affect gene expression. The enzyme activity associated with alcohol dehydrogenase (ADH), which is produced in the liver, breaks down alcohol. Some people, such as Northeast Asians and Native Americans, produce less ADH and as a result become intoxicated more quickly than others. Women in general also produce less ADH than do men. By drinking alcohol on a regular basis, Matt produced changes in gene expression to make more ADH. Regular drinking prompted cells in his liver to increase copies of the *ADH* gene, by reading the *ADH* gene more efficiently. Had he maintained sobriety for an extended period of time, his liver would have produced less ADH. These epigenetic mechanisms for tolerance made it possible for him to drink often without as much of an intoxicating effect as a person who drinks less often. (Addiction and habit formation issues are discussed in Chapter 6).

The key point here is that epigenetic effects go beyond drug-gene-cell interactions. How people mentally frame their expectations also can significantly contribute to the feedback loop between drug-gene interactions.

Developmental Genomics

Epigenetic effects can also result from social factors. The bidirectional relationship between social factors and gene expression is explored in the field of social genomics (Cole et al., 2007). Ernie Rossi (2002) has speculated since the late 1980s that psychotherapy could play a role in gene expression. Developmental and social factors play a significant role in gene expression, and researchers have begun to identify the types of genes subject to social regulation and the biological signaling pathways mediating those effects.

When psychodynamic theorists proposed that "good-enough" mothering, and when attachment theorists showed that secure attachment resulted in good mental health later in life, they had no inkling that epigenetic effects were involved—it was beyond the science of the era. Now we can appreciate from the perspective of the mind-brain-gene feedback loops that people who were well nurtured tend to become more resilient later in life. In contrast, people who were not tend to be more easily stressed and prone to develop multiple health and other psychological problems, such as anxiety and depression. Good nurturing stimulates the release of serotonin, which aids in the expression of enzymes in the hippocampus. These epigenetic processes result in decreased DNA methylation of the cortisol receptor gene, enhancing the production of cortisol receptors. Ultimately this negative feedback loop provides a thermostat for stress and increases resilience (Weaver et al., 2004).

Much of the pioneering work on these epigenetic effects was performed by Michael Meaney and colleagues of McGill University, initially on rat mothers and their pups (Francis & Meaney, 1999). He and his colleagues went on to examine the brains of people that committed suicide by analyzing the levels of DNA methylation at the cortisol receptor gene in their hippocampus. They found that DNA methylation tended to be higher in people who had a history of early childhood abuse or neglect. By contrast, the DNA methylation levels at this gene were relatively low in the suicide victims who did not have a traumatic childhood. The high methylation levels in the abuse victims drove down expression of the cortisol receptor gene, making the negative feedback loop less efficient and allowing levels of cortisol to stay high. Consequently, these people tend to be more easily stressed. Others have also found that the brains of adults who died of suicide and had a history of ACEs exhibited epigenetic changes to the cortisol receptor gene (McGowan et al., 2009). Of course, it isn't just people that experience ACEs who later become anxious, depressed, or suicidal. But ACEs do increase vulnerability to becoming anxious and depressed, as well as raising the risk for suicide, through complex interactions with later psychological distress—as the risk factors increase, so does the potential for self-harm behaviors.

People who commit suicide who had a cross-generation family history of neglect also show low opioid receptor activity in the brain (Gross-Isseroff, Biegon, Voet, & Wezman, 1998). As explored in Chapter 2, attachment can play a significant role in the development of the capacity to self-sooth. The opioid system can be augmented later in life through positive relationships, including psychotherapy.

The ACE study showed a significant correspondence between ACE score and depression, as well as suicide. Just as the attachment theorists predicted, children with insecure attachment, as well as neglected or abused children, such as Gina, suffer from underdevelopment in stress tolerance. The failure to turn off the stress response occurs at many levels in the mind-brain-gene feedback loops. (The stress response systems is explored in Chapter 6.) One of the stress response subsystems, referred to as the neuroendrocrine system, overactivated when Gina encountered stress. When she detected even a slight threat, her amygdala activated her hypothalamus to release corticotropin-releasing hormone (CRH) and arginine vasopressin. These two neurohormones stimulated her pituitary gland, which responded by releasing adrenocorticotropic hormone (ACTH) into her bloodstream. Cells in her adrenals took up this hormone and then released cortisol, which provided a systemic boost to her energy by increasing glucose availability and dopamine release. This complex system is the hypothalamic-pituitary-adrenal (HPA) axis, a major stress response system operating in the mind-brain-body. Those interactions are normal for realistic short-term threats, but what was not normal was that even after a presumed threat was long gone, her stress system stayed activated.

In contrast, Sara's stress response system worked well, with negative feedback mechanisms in place to shut it off during periods when there was no longer any danger. Part of Sara's negative feedback loop included cortisol receptors on her hippocampus that acted like a thermostat: when the level of cortisol reached a certain level, those receptors triggered events that signaled her hypothalamus to turn down the activity in the HPA axis. Sara's secure attachment had built in the thermostatic infrastructure of resiliency and durability. In contrast, Gina's thermostat system was underdeveloped, which kept her needlessly on edge.

Despite being identical twins, Gina and Sara differed in their durability to stress. The differences were programmed epigenetically by secure attachment for Sara and undermined by ACEs for Gina. With Gina, trauma and neglect played a significant role in epigenetically programming her brain by preparing her for more adversity. To add insult to injury, this heightened state of vigilance to potential danger came at a cost, making Gina vulnerable to anxiety, depression, addiction to comfort foods, and health problems years after the adversity. When she encountered even mild levels of stress, her cortisol levels remained excessive, potentially shrinking her hippocampus. With an impaired hippocampus she was less able to construct new explicit memories and instead repeated negativistic episodic memories during her extended default-mode network periods.

In contrast, Sara's secure attachment epigenetically endowed her with many cortisol receptors on her hippocampus that functioned as part of a negative feedback loop to turn down the release of cortisol when there is no need for continued stress response. Overall, Sara was adaptable and resilient. Not so with Gina, who was easily stressed and not adaptable to new situations. With both of them, we see that experience changed gene expression, which then fed back to affect their subsequent experiences.

> As the infant experiences nurturing, her serotonin levels increase, which signals her hippocampus to increase the production of an enzyme that acetylates histones. This enzyme binds to the cortisol receptor gene and adds acetyl groups to histone proteins. The histone acetylation creates a more relaxed environment so that DNA methylation is removed. The decreased DNA methylation leads to higher

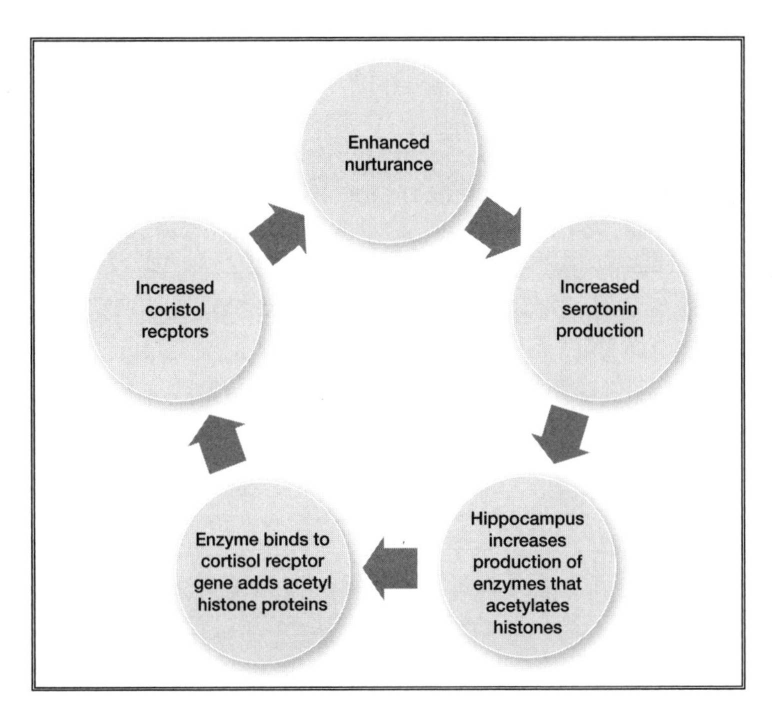

Figure 3.1: The cycle of the epigenetic development of stress tolerance provided by secure attachment.

expression of the cortisol receptor gene and consequently better ability to turn off the HPA axis—the stress response system.

Maternal separation, in contrast, leads to decreased DNA methylation of the arginine vaseopressin gene. This results in the increased production of arginine vaseopressin, which stimulates the HPA axis—the stress response system. Meanwhile, because of the methylation of the cortisol receptor gene, it is difficult to turn off the stress response.

One of the first studies that highlighted the relationship between social factors and gene expression examined the extent to which individuals felt socially connected to others (Cole et al., 2007). Those people who felt lonely and distant from others expressed genes that selectively upregulated inflammation, while genes involved in responses to viral infections and production of antibodies by B lymphocytes were downregulated. The key factor in the mind-brain-gene feedback loops was that the psychological pathways mediating gene expression were linked to a person's subjective sense of isolation, rather than their objective number of social contacts.

A variety of studies have identified genetic transcription correlates of other socioenvironmental conditions. For example, those from backgrounds of low socioeconomic status tend to interpret ambiguous situations as threatening, and that perception is more strongly linked to differences in gene expression (Chen et al., 2007). Similarly, people experiencing chronic threat of loss, such as having a spouse with cancer, tend to upregulate the expression of leukocyte inflammatory genes (Miller et al., 2008).

Transgenerational Trauma

My paternal grandmother, Helen, barely escaped the Armenian genocide after her husband was killed. Together with her four-year-old daughter, Alice, and her brothers, they were herded out with thousands of others in what was later called the death march. Many thousands were killed, raped, or maimed. Helen was subsequently separated from Alice for eight years. By the time they were reunited, Helen had remarried and had three more sons. The oldest was my father. In terms of any potential transgenerational epigenetic effects, only one family member—one of my two uncles—developed problems: he died of alcoholism-related causes, but with no apparent anxiety or depression.

In contrast, my maternal grandmother, Louise, in one month's time got married to my grandfather, whom she had known for just a few days, and fled their home town to escape the genocide. Much of the following two years, she and her husband moved from country to country in hopes of finding a US consulate that would allow immigration. While they were searching for a safe place to live, Louise became pregnant. Their baby, Marjorie, was born one year before they were finally granted visas travel to the United States. During that anxiety-provoking year of waiting they bonded with Marjorie. She was my mother.

How did they and almost all their offspring not only survive but later thrive? What could optimally provide resilience in response to stress could flip to stress intolerance via epigenetic mechanisms. Secure attachment offered by Helen and Louise and the support of an extended family and community provided support for the development of a resiliency infrastructure, as well as buffer the effects of trauma. They were some of the most generous and warm people I have ever met.

On the other hand, Gina's adoptive mother, Sonya, endured ACEs before the war in Bosnia, survived it, and then later was the victim of domestic violence. People such as Sonya who endured ACEs and subsequent trauma had higher than normal levels of CRH, ACTH, and cortisol resulting in low stress tolerance and higher anxiety and depression. Her depression blunted her ability to nurture Gina. The epigenetic effects on Gina resulted in similar neurochemical dysregulations as Sonya despite not sharing the same genes.

What do we now know about pregnancy during stressful and traumatic periods? Had Sonya been pregnant at the time of the Bosnian genocide, her child could probably have been born with a stress system primed for anxiety and depression. The effects of trauma experienced by mothers may be passed on biologically to their offspring, especially if the mothers were pregnant. Mothers who were pregnant at the time of the World Trade Center attacks gave birth to children with elevated stress response, especially a hypersensitive HPA axis (Yehuda et al., 2005). There are many potential variations of these transgenerational effects of stress and trauma. On the most obvious level, mothers and babies, like Louise and Marjorie, shared a blood supply that was marked by increased cortisol, contributing to a ramped up stress system and possibility vulnerability to anxiety, PTSD, and depression. Fortunately for Marjorie, and me for that matter, she did not encounter later trauma. Secure attachment and a warm supportive extended family buffered much of the stress effect. However, there have been reports that high levels of stress hormones during pregnancy contribute to giving birth to babies who may be epigenetically programmed for risk of affective disorders (Seckl, 2008).

The offspring of mothers who suffered PTSD associated with the Holocaust tend to be more prone to develop PTSD, even though they were born long after the Holocaust (Yehuda & Bierer, 2007). Though children of Holocaust survivors are more prone to depression, those in the second generation tend to be more vulnerable to PTSD if their mothers suffered PTSD. These potential epigenetic effects appear to occur through maternal inheritance. No correlation has been found for children whose fathers experienced PTSD as a result of the Holocaust.

The Neurochemistry of Resilience

A variety of neurotrophic factors promote resiliency that appear to be activated through epigenetic mechanisms. Brain-derived neurotrophic factor (BDNF) has been called one of the "brain fertilizers" because it has a significant role in neurogenesis, neuroplasticity, and overall brain health. Diet and exercise promote the release of BDNF (explored in Chapter 5). It also appears that the behavior we call play induces *BDNF* gene expression (Panksepp & Biven, 2012; Gordon, Burke, Akil, Watson, & Panksepp, 2003). Another neurotrophic factor aroused by play, insulin-like growth factor, is associated with brain growth and maturation and can be epigenetically facilitated by prosocial circuits in the brain (Burgdorf, Kroes, Beinfeld, Panksepp, & Moskal, 2010). Both Helen and Alice expressed warmth and nurturing in playfulness. These epigenetic dynamics support the idea that therapy can be enhanced by periodic playfulness.

The expression of various neurotrophic factors in the nucleus accumbens are associated with greater resilience. For example, decreased methylation of the glial cell-derived neurotrophic factor (GDNF) gene is associated with greater stress tolerance. In contrast, increased methylation of the GDNF gene is associated with decreased stress tolerance. GDNF is essential for regulating dopamine release in the nucleus accumbens and for the survival of dopaminergic neurons. Chronic unpredictable stress alters dopaminergic function, so that dopaminergic circuits reshape in the prefrontal cortex, impairing decision making and inhibiting maladaptive habits and working memory.

Another key neurotrophic factor epigenetically expressed in those who are well nurtured involves the transcription factor called nerve growth factor (NGF), which binds to the cortisol receptor gene to increase cortisol receptors. However, when the cortisol receptor is suppressed, as occurs with neglect or child abuse, NGF does not bind well and fewer cortisol receptors are produced in the hippocampus (McGowan et al., 2009). When this "thermostat" is impaired, the stress axis is difficult to turn off, predisposing a person like Gina to anxiety and fearfulness.

Women such as Gina who were neglected as babies by their mothers tend to be neglectful mothers themselves through the alteration of the *NGF* gene (Champagne & Curley, 2009). They tend to have offspring with the same alteration to their *NGF* genes and so become stressed out mothers too, continuing the cycle of neglect. Sonya had not experienced abuse and neglect as a child; had Sonya been Gina's biological mother, Gina may not have initially carried an alteration of her *NGF* gene. However, because of her adoptive parents' neglect and abuse, she would have acquired the alteration of the *NGF* gene through experience. These transgenerational effects are dynamic: poor parenting perpetuates poor parenting, while good parenting perpetuates good parenting.

Epigenetic changes also occur to neurohormones such as estrogen and oxytocin. Normally, after giving birth these neurohormones play a major role in promoting bonding. However, in women who did not received adequate mothering themselves, levels of estrogen receptors are reduced (Champagne, et al., 2006). Low estrogen reduces the binding of oxytocin in the hypothalamus. Oxytocin promotes maternal and prosocial behavior, which fails to develop adequately in response to poor parenting. On the other hand, good parenting increases estrogen receptors, which stimulates the expression of the oxytocin receptor gene in the hypothalamus and so promotes bonding. The bonding between Helen and her four children promoted resiliency as a family system. Only my one uncle, who died of alcoholism, suffered.

While complex feedback loops between genes, neurophysiology, and attachment dynamics can persist, they may be potentially malleable given significant changes in life experiences. For example, individuals born to mothers who received good nurturing from their own mothers but who were adopted and raised by poorly nurturing grandmothers have lower levels of the estrogen receptor in their hypothalamus than siblings who remained with their biological mother (Champagne et al., 2006). The reverse occurs for those born to poor nurturers but raised by good nurturers: they have elevated levels of the estrogen receptor in the hypothalamus and became good mothers.

Enhanced nurturing promotes resiliency through changes such as diminished levels of stress neurochemistry (e.g., CRH and ACTH) and development of more GABA receptor sites, which all act to reduce stress and anxiety (Lee et al., 2005; Carpenter et al., 2004). These neurochemical changes generally promote greater calmness. Similarly, those who experience optimal nurturing and/or psychotherapy tend to be more focused and energized, as evidenced by the development of more receptors for glutamate and norepinephrine, facilitating learning and adaptability. Overall, epigenetic and attachment research converges to support the theory that well-nurtured individuals are less anxious and more engaged in their environments, even during fear-inducting situations (Zhang & Meaney, 2010).

Telomeres, Health, and Longevity

Another way that our behavior and environment may change our genome and its interactions is by the health of our chromosomes. Telomeres are the DNA complexes (a.k.a. "junk DNA") that cap and protect the ends of the chromosomes, like the plastic ends of shoe laces; as each cell divides, telomeres shorten. The availability of an enzyme that adds nucleotides to the telomeres, called telomerase, has been linked to health and psychological factors, Telomere length serves as a biomarker for health and longevity, and longer telomeres promote chromosomal stability and health; our poor self-care behavior can accelerate shortening of our telomeres and thus impair our chromosomes. Serious illnesses, such as cardiovascular disease, atherosclerosis, myocardial infarction, vascular dementia, hypertension, diabetes, and obesity, shorten telomere length. There has been accumulating evidence that accelerated aging is associated with unremitting stress as measured by the shortening of telomeres. For example, chronic stress experienced when a person cares for a terminally ill family member has been associated with a reduction in telomere length of the caregiver (Epel et al., 2004).

Depression and anxiety have been linked to shorter telomeres. The more severe the depression and/or anxiety, the shorter the telomeres (Blackburn & Epel, 2017). Chronic and perceived stress is associated with shorter telomeres as well as telomerase. Women with the highest levels of perceived stress have been found to have the shortest telomeres, shorter on average by the equivalent of at least one decade of additional aging (Epel et al., 2004).

A study of approximately 12,000 Chinese women found that depression was associated with shorter telomeres, that the more severe and prolonged the depression the shorter they were (Cai et al., 2015), and that the greater the childhood adversity and subsequent depression, the shorter the telomeres. A study in the Netherlands of approximately 3,000 people found that those who suffered from depression lasting less than ten months did not have significantly shorter telomeres, but those with depression for more than ten months did (Verhoeven et al., 2014). Just as depression shortens telomere length, short telomeres may contribute to depression. This is especially the case if there are shorter telomeres on chromosomes in the hippocampus (Mamdani et al., 2015).

Overall, poor self-care practices, illness, and psychological disorders shorten telomeres. And shorter telomeres contribute to illness and psychological disorders. This makes the role of psychotherapy in promoting behavioral health significant with respect to promoting telomere length and thus mental health.

> Telomerase is an enzyme that adds nucleotides to and protects telomeric ends. Insulin growth factor, vascular endothelial growth factor, and epidermal growth factor all upregulate telomerase activity, while transforming growth factor beta inhibits it (Lin, Epel, & Blackburn, 2009). Oxidative stress reduces telomerase activity, whereas antioxidants increase it. Factors that enhance brain health, encoded in the mnemonic SEEDS—social, exercise, education, diet, and sleep (described in Chapter 5)—all protect telomeres, and their absence shortens telomeres. In addition, telomerase-induced lengthening is produced during meditation (Jacobs et al., 2011).

Whether involving gene expression, suppression, or telomere length, there are multidimensional feedback loops within the mind-brain-gene continuum. Our genes, chromosomes, and cells are all dramatically affected by our lifestyle, stress level, exercise, diet, sleep patterns, and how our mind chooses to orchestrate all these factors. Our environment, behavior, and mental networks interact to play significant roles in changing our biology down to gene expression. Because psychotherapy seeks to enhance behavioral health, it promotes positive gene expression and telomere length.

Genetic reductionism has given way to a far more dynamic new vision of mind-brain-gene feedback loops. It is no longer genes causing psychological disorders; it is the interaction of these dimensions that we must understand so we can truly help people who are suffering. Mind-brain-immune feedback effects are also dynamic—that poor health perpetuates poor stress tolerance and depression, while good health increases potential resiliency and better health, as explored in Chapter 4.

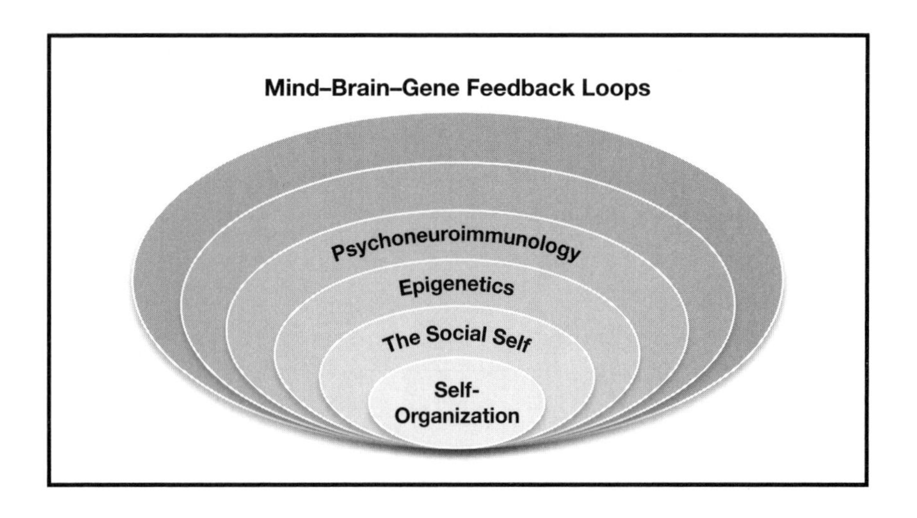
Mind–Brain–Gene Feedback Loops
Psychoneuroimmunology
Epigenetics
The Social Self
Self-
Organization

FOUR

The Body-Mind and Health

Think with your whole body.

—Taisen Deshimaru

Anna and Michael had been married for seventeen years when both began to complain that the other no longer focused on the rest of the family. They spent an inordinate amount of time ruminating resentments about each other. Meanwhile, their two children were entering the first few years of high school and presenting new challenges for their parents. Both Anna and Michael felt they didn't have the energy to keep up with the constant attention needed to maintain clear limits and expectations for the kids. Though they did not want to admit it, they felt relieved when their kids began to spend more time on their computers, playing video games and on social media. This meant less monitoring was necessary because the kids left the house less, but they began to match their parents in obesity, fatigue, and dysphoric moods.

Perplexed by everyone's loss of energy, Anna asked their physician whether the entire family had contracted Lyme disease. They felt ill and did not know why. He ordered blood tests for each of the family members. Though there was no evidence of Lyme disease, he expressed concern that they all had become significantly overweight. He also reported that both Anna and Michael had high levels of C-reactive protein (a measure of inflammation), blood glucose, and LDL (bad) cholesterol. Anna had developed type 2 diabetes, and Michael had metabolic syndrome (a cluster of conditions—increased blood pressure, high blood sugar, excess body fat around the waist, and abnormal cholesterol or triglyceride levels that occur together, increasing the risk of heart disease, stroke and diabetes). Anna responded by saying, "We were already very depressed! Now you are telling us we have bad genes? That makes me even feel worse." Michael agreed. In response, their physician prescribed Prozac for both of them. With this medication Band-Aid their physician missed the opportunity to offer comprehensive health care and refer them to therapists to avert disastrous long-term mental and physical health. Though he started the consultation constructively by warning the entire family about their weight and both parents of their looming illnesses, the integrative approach they needed was compromised by the quick fix of "mismanaged" care.

What role should psychotherapy play in helping this family? Psychotherapy in the twenty-first century could be renamed "behavioral health," because self-care behaviors have major effects on the immune system, the brain, and the body in general. These interactions have a profound effect on mental health. It is this relationship that is explored in this chapter.

BEHAVIORAL HEALTH

Anna and Michael's family has become the new norm. There are now overwhelming numbers of people like them throughout the developed world. Plagued with health problems brought on by poor physical and emotional self-care, they suffer bidirectional causal pathways between acquired physical and psychological impairments.

The Centers for Disease Control estimates that health behaviors account for 50 percent of adverse health outcomes in the United States—as much as genetics (20 percent), the environment (20 percent), and access to health care (10 percent) combined. These statistics suggest that half of all health conditions are preventable by changes in self-care behavior (Amara et al., 2003). Research has found that 40 percent of medical patients have a comorbid psychological disorder, while 75 percent of patients with a psychological disorder also suffer with a comorbid physical disorder (Kessler, Ormel, Demler, & Stang, 2003). Essentially, health behaviors represent the interwoven natures of physical and mental health.

The number of Americans suffering at least one physical illness is predicted to increase to 157 million by the year 2020. This is a mental health crisis that can no longer be overlooked. Though these statistics are ominous, integrated health care providers, including psychotherapists, can work to avert disaster to the health of millions of people through a better understanding of the bidirectional causal interactions among the mind, brain, and the immune system found in the field of psychoneuroimmunology, and providing approaches consistent with it. Since its emergence as a rigorous field of research, psychoneuroimmunology has identified many interrelated mental and physical health dysregulations. Not so coincidentally, this field of inquiry the emerged with the surge in numbers of people, like Anna and Michael, with chronic and acquired illnesses that dysregulate the immune system who also suffer from psychological disorders.

> Because over half the population of the United States unknowingly suffers from self-inflicted immune system dysregulation, psychotherapy in the twenty-first century must promote lifestyle and behavioral health changes. Not doing so is like building a house on a sandbar of a hurricane-swept beach.

THE IMMUNE SYSTEM

To get a better idea how chronic health conditions develop and cause psychological disorders, it is useful to put the immune system in perspective. Just as we are optimally endowed with a stress response system to deal with external danger, such as fighting off or fleeing from a predator, our bodies are protected from pathogens by a dynamic immune system. Whether the threat is from foreign bacteria, a contagious virus, or simply a cut on the finger, your body marshals internal resources to protect its cells and maintain homeostasis. Like the police, fire, and ambulance services combined, the immune system protects the body from external and internal threat. When working optimally it can save your life. When activated inappropriately it can cripple your life.

As a diffuse sense organ scattered throughout the body, the immune system communicates to the brain by both neural (fast) and hormone (slower) subsystems, influencing mood, cognition, and behavior. It comprises two main components: specialized cells that carry out protective functions and chemical messengers that allow those cells to communicate with one another and the rest of the body. The cells and the chemical messengers interact to mediate the location and intensity of inflammatory responses to protect the body from harm.

Chronic stress combined with poor self-care, such as inadequate sleep, impoverished diet, lack of exercise, and extra weight, inappropriately activates the immune system, with damaging effects. Anna and Michael, like millions of other people, acquired chronic conditions that turned their dysregulated immune systems into threats. Their immune systems switched from protectors to overactive enemies triggering autoimmune disorders and a downward spiral of significantly compromised physical and emotional health. Whether in response to adverse childhood experiences or chronic stress, or simply because of poor self-care, autoimmune disorders result in a variety psychological disorders, which then further exacerbate existing autoimmune disorders. To understand how this occurs and how psychotherapists can intervene, it is useful to highlight the multiple feedback loops that make up the immune system.

> Specialized cells that carry out protective functions include the lymphocytes, which come in a wide variety of cell types, including B and T cells, produced in the bone marrow and thymus gland, respectively. Macrophages are the general foot soldiers that gobble up threatening bacteria, memory B cells are the snipers trained to attack specific targets, and helper T cells are communication officers, alerting other troops to an invasion. To maintain homeostasis in the body, these cells work together in a precise and coordinated dance choreographed by

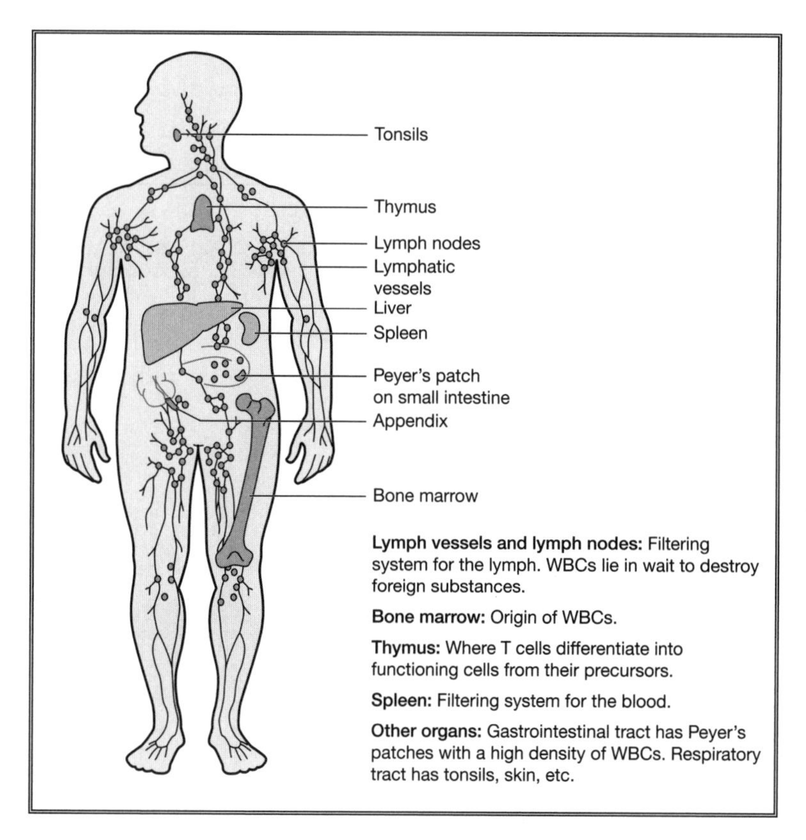

Figure 4.1: The components of the immune sytem.

chemokines and cytokines, chemical messengers that allow those cells to communicate with one another and the rest of the body.

After lymphocytes are born they are tested in the bone marrow and thymus to be exposed to molecules of your body so that they will not attack the body. The ones that bind strongly to "self" molecules are killed off or undergo "editing" in the genes that give rise to receptors. Autoimmune disorders develop when this process fails. Facilitated through cytokines, pathogens are killed while avoiding harm to the body.

When harmful substances enter our body, cells in our immune system called macrophages (big eaters) serve as an immediate defense.

These Pac-Men of the immune system detect, destroy, and clear foreign substances from the body. This first response system also includes neutrophils (40–75 percent of white blood cells), which aid the macrophages in killing foreign substances while initiating the inflammatory response. The term *inflammation* comes from a Latin term meaning "set on fire"; inflammation serves as a protective mechanism so that the healing processes of tissue repair can begin. For local responses, like a bruise on the knee, inflammation also produces a physical barrier that prevents the spread of infection into the blood stream.

Chronic Inflammation

While short-term inflammation that responds to injury or illness represents a healthy process, chronic inflammation is not. Chronic inflammation represents a common factor among many psychological disorders and poor health. Because it is such a dominant feature, a better understanding of chronic inflammation will help our efforts to put it under control. In other words, to understand how a healthy system becomes unhealthy so that we can shift it back to healthy again, we need to understand how chronic inflammation gets turned on and off.

Normally, as part of the inflammatory response, infected or damaged cells send out alarm signals via chemical messengers called cytokines that attract and guide specific immune cells to the site of infection or damage. Cytokines (*cyto* means "cell" in Greek, and *kinos* means "movement between cells") are communication substances, proteins released by immune cells that act on target cells to regulate immunity. Cytokines include the interleukins (meaning "between the white blood cells") and tumor necrosis factor alpha. There are pro-inflammatory and anti-inflammatory cytokines. The pro-inflammatory cytokines (PICs) coordinate inflammatory responses. Anti-inflammatory cytokines, as their name implies, work to dampen the inflammatory response. With chronic inflammation the PICs dominate. Inadequate diet, poor-quality sleep, and lack of exercise (the topics of Chapter 5), as well as stress, depression, autoimmune disorders, and obesity, are associated with excess release of PICs, and thus chronic inflammation.As a common marker of inflammation, a fluid produced by the liver called C-reactive protein (CRP) can be measured in a standard blood panel. One of the findings from Anna's and Michael's blood panels was moderately high levels of CRP. While very high levels of CRP are caused by infections, moderately high levels are associated with autoimmune diseases, obesity, and depression (Kendall-Tackett, 2010). Chronic inflammation represents one of the common factors associated with autoimmune disorders, such as rheumatoid arthritis, lupus, type 2 diabetes, Addison's disease, Crohn's disease, and ulcerative colitis. Chronic inflammation is also associated with neurodegenerative diseases such as multiple sclerosis, Parkinson's disease, and Alzheimer's disease, and with psychological disorders.

Inflammation and Associated Psychological Disorders

- Depression
- Bipolar disorder
- Autism
- Posttraumatic stress disorder
- Cognitive decline

In addition, in children inflammation has been linked to:

- Tourette's syndrome
- Obsessive compulsive disorder
- Attention deficit hyperactivity disorder

Anna and Michael, along with their kids, were unknowingly stoking up the amount of PICs and offering themselves few opportunities to generate anti-inflammatory cytokines through life style changes (described in Chapter 5). As a result, they were all suffering from chronic inflammation, with its associated fatigue, cognitive deficits, and depression.

When Anna returned to her family doctor for follow-up, complaining that she felt too little improvement, she was referred to me. Initially, she was quite perturbed that I addressed her poor self-care behaviors and associated them with her depression. Her resistance to these factors began to dissipate when I noted that both of her children began developing some of the same symptoms when their self-care matched hers.

The Brain's Immune System

Chronic inflammation can lead to depression, cognitive impairments, fatigue, and achiness. Brain systems are affected by inflammation that can also perpetuate inflammation. These systems directly affect energy levels, pain, coping responses, and sense of self. So when Anna and Michael complained about lethargy, difficulty generating positive thoughts, depression, and cognitive deficits, inflammation likely played a significant role.

The brain has specialized cells with immune-like functions. Chief among them are the glial cells. Initially glial cells were thought of as the substance that holds neurons together (*glia* means "glue") and were erroneously thought to function only as support cells. Glial cells include microglia and astrocytes, which are considered part of the brain's resident immune cells.

Microglia make up 6–12 percent of all the cells in the central nervous system, where they constantly monitor for potential immune problems. They have many of the same receptors as peripheral macrophages and so can recognize bacteria and viruses. When they detect danger microglia release PICs such as interleukin-1 (IL-1) (Maier & Watkins, 2009). Chronic stress and compromised health "prime" microglia so that they make and release PICs more easily when they encounter danger again. In other words, once the immune cells in the brain have been activated, it is more likely that activation will occur in the future.

Astrocytes play a significant role in the immune system by providing a point of interaction between cytokines and neurons via genetic transcription and synaptic plasticity. Astrocytes exchange signals with neurons, detect and react to immune signals, and release PICs, which influence peripheral immune cells. Through this process of monitoring, reaction, and learning, astrocytes can play significant roles in the perpetuation of the inflammatory spiral.

Activation of the inflammatory pathways in the brain adversely affect memory and mood. The excessive release of PICs from microglia and astrocytes in the brain, as well as in the rest of the body, including from fat cells, causes wide-ranging detrimental psychological effects. Overexpressed PICs cause cognitive deficits that involve disturbances in synaptic strength. High concentrations of receptors for PICs are located in the prefrontal cortex and hippocampus, potentiating cognitive impairments, including poorer working memory, episodic memory, and execu-

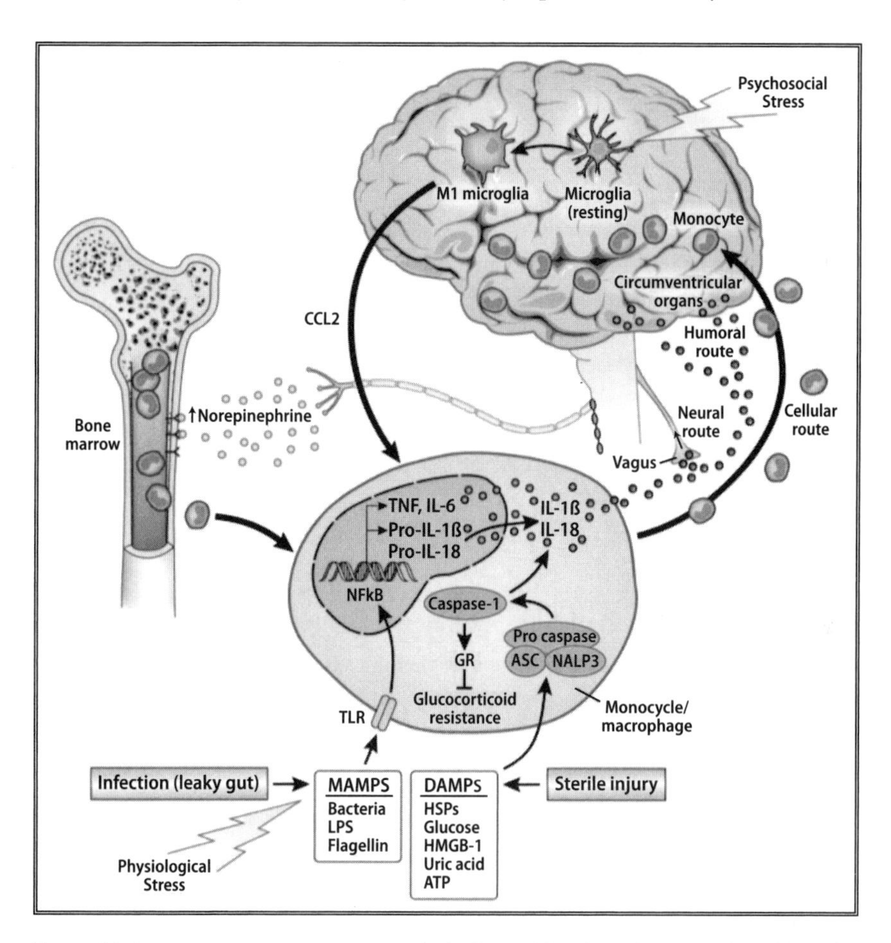

Figure 4.2: Immune system interactions with the brain. Note how the microglia cells in the brain function as the resident immune system of the brain. Microglial activation results in the production of pro-inflammatory cytokines such as interleukin (IL)-1β, (IL-6), and tumor necrosis factor-α (TNF-α) influencing the surrounding brain tissue.

tive functions (Lin & Marsland, 2014). This is why Annie and Michael had an executive network "brownout." For example, excessive IL-1 in the hippocampus has been shown to impair memory by interfering with brain-derived neurotrophic factor (BDNF), which is involved in neural plasticity, neurogenesis, memory, energy balance, and mood.

> Because chronic inflammation deteriorates overall health and contributes to cognitive deficits and mood disorders, promoting lifestyle changes that lower inflammation should be a major goal of therapy. For example, given that extra fat cells contribute to chronic inflammation, weight loss should be a major goal—but this also represents a major challenge to building the therapeutic alliance!

The Mental Health Consequences of Excessive Fat Cells

According to the World Health Organization approximately 2 billion people are overweight. Approximately 68 percent of the population of the United States is overweight. As of 2016, 35 percent of men, 40 percent of women, and 17 percent of children and adolescents were obese—the United States leads the world with this pandemic. Obesity contributes to many other chronic health conditions strongly associated with psychological disorders. On average, obese people over 40 will die 6 to 7 years earlier than non-obese people. And those fewer years are marked by cognitive deficits and emotional dysregulation. Obesity shortens telomeres more than smoking.

Being overweight or obese causes a systemic feedforward breakdown in multiple homeostatic pathways, including a condition referred to as leptin resistance, further fueling obesity and rising cortisol levels (Newcomber et al., 1998). Obesity also causes inflammation, which in turn increases the risk of illnesses like autoimmune disorders, coronary heart disease, stroke, hypertension, sleep apnea, and type 2 diabetes. These physical illnesses are all associated with depression, anxiety, and cognitive impairments.

The Centers for Disease Control and Prevention defines obesity as a body mass index (BMI) over 30. Independent of age, BMI is inversely associated with a range of cognitive impairments and temporal lobe and global brain atrophy, cognitive decline, and incidence of dementia. A meta-analysis showed that people who are obese during midlife harbor a 1.35-fold increase risk for dementia late in life (Anstey, Cherbuin, Budge, & Young, 2011). A large study spanning thirty-six years found that those who were obese in midlife were three times more likely to develop Alzheimer's disease than those who were not obese (Whitmer et al., 2005). The higher the BMI, the greater the circulating IL-1. And where the fat cells were located is critically significant. Central adiposity (belly fat) predicts dementia independent of BMI. In fact, people with normal weight who nonetheless had central adiposity were 89 percent more likely to develop dementia than people without it (Whitmer et al., 2008). In other words, the larger the belly, the greater the risk of dementia. These findings have led to the concept that an apple-shaped body places you at a greater risk than a pear shape. Because Michael carried an abnormally large belly, he was at greater risk for cognitive, mood problems, and dementia.

Obesity blocks BDNF, which is one of the many ways it contributes to dementia and depression. With the increase in PICs associated with obesity, there is a corresponding decrease in BDNF. Obesity also results in the dysregulation of energy intake and expenditure. Because BDNF plays an important role in energy balance, it can reduce inappropriate feeding while increasing energy output. Obesity increases feeding and decreases energy.

With the pandemic of overweight and obese people in the United States, the number of people with type 2 diabetes (90–95 percent of all those with diabetes) is well over 28 million (Nezu, Raggio, Evans, & Nezu, 2013). By far the most significant risk factor of type 2 diabetes is obesity, and especially fat above the hips. This makes excessive belly fat diagnostic of not only health problems but also cognitive and mood problems.

To understand the relationship between obesity and immune system dysregulation, it is useful to consider that obesity is regarded as a chronic subclinical inflammatory condition. Not only do adipose tissues (fat cells) swell up, but the dead cells are not cleared out efficiently in obese individuals, setting in motion inflammatory cells and macrophages to cluster around dead and degenerated cells to engulf and digest them. Like a stagnant pond with muck building up because there are no streams coming in or out, in fatty tissues fat cells decay without any clearing or nourishment. Chronic inflammation present in obese people is distinct from acute inflammation—it is no longer involved in tissue repair. Fat cells leach out PICs, producing 10–35 percent more circulating PICs, such as IL-6, than in nonobese people. This makes these cells function as agents of mood instability and cognitive impairment.

In addition to contributing to depression and cognitive deficits, IL-6 is associated with coronary heart disease and insulin resistance and is increased in people who are obese (Brunn et al., 2003). The higher the IL-6, the greater the risk of developing the type 2 diabetes. And with less glucose getting into cells, the brain is starved of fuel. In fact, some neurologists are calling Alzheimer's disease "diabetes type 3."

Autoimmune and Psychological Disorders

Inflammation represents the common denominator for metabolic syndrome, obesity, and depression. These conditions break down many of the homeostatic processes important to maintain general health and mental health. Though the periodic inflammatory response is critical for tissue repair and homeostasis, chronic inflammation dysregulates the immune system feedback loops, leading to a variety of pathological conditions, such as autoimmune disorders.

Like Michael, one in five Americans likely have metabolic syndrome. The percentage of people increases with age, reaching more than 40 percent among people in their sixties and seventies (Resnick & Howard, 2002). Metabolic syndrome is marked by insulin resistance, dyslipidemia (abnormally elevated cholesterol fats in the blood), and elevated blood pressure. All these factors are strongly predictive of cardiovascular disease and accelerated neurocognitive deficits.

Both a symptom and cause of major health problems and psychological disorders, autoimmune diseases represent bidirectional causal interactions with psychological disorders. For example, it is now well documented not only that people who develop type 2 diabetes are prone to depression. Once the inflammatory spiral begins, it is difficult to put on the brakes.

Anna would have been best served had their family physician recommended significant lifestyle changes to pull out of the self-perpetuating nose dive into inflammation, diabetes, depression, and cognitive deficits.

People with type 2 diabetes are twice as likely to experience depression as their counterparts who do not suffer from diabetes. Large meta-analytic studies have shown that there is a bidirectional causal relationship between type 2 diabetes and depression across diverse populations (Nezu et al., 2013). In other words, type 2 diabetes increases the likelihood of becoming depressed, and depression increases the likelihood of developing type 2 diabetes.

The Stress-Inflammation Connection

Michael complained that when he began to feel overwhelmed with stress he seemed to pick up any virus going around. Eventually, he transitioned into a phase of feeling like he had the mild case of the flu all the time: achy and tired, with a queasy stomach, but never breaking out into a full blown flu with a fever and its associated symptoms.

It has long been the folklore of psychology that stress dampens the immune system. Short-term stress, in fact, has been shown to suppress T cell formation and natural killer cell function and to decrease the ability to repair broken DNA. But stress can enhance inflammation, too. Stress can activate PICs, including within the glial cells in the brain.

Michael's chronic stress felt like an odd mix of feeling nervous and fatigued at the same time. Chronic stress can cause epigenetic changes in the expression of PIC genes in immune cells. With more peripheral inflammation Michael experienced, the more neuroinflammation he had. With mood and cognition compromised, his self-care deteriorated. He joined Anna in eating more and moving less. With significant weight gain, he also joined her in malaise and obesity.

Addressing Anna's poor self-care, associated health problems, and depression was only the beginning of therapy. There was a backstory that Anna did not tell her physician: she had experienced increasing stress during the previous few years. Her boss demoted her after her job performance had begun to suffer because of low energy, and her new position required that she work swing shifts. Because her employer did not provide adequate parking, finding her car on a dark street at midnight presented a stressful challenge. Because the neighborhood was crime infested, she asked that the security guards provide an escort to her car. After her boss told the security department not to allow any of the guards to leave company property, one night she was physically assaulted and robbed. In response to her union grievance, her boss relented and allowed a security guard to escort to her car, but the residual symptoms of the trauma persisted. She found little support at home during the year before visiting her primary care physician. The only semblance of comfort she found was "vegging out" on the couch watching movies with Michael.

Anna's chronic stress, exacerbated by trauma and poor self-care, combined to trigger an autoimmune disorder. Multiple physiological pathways contributed to her increased inflammation and decreased stress tolerance. The chronic and acute stress disrupted the insulin balance in her body and led to insulin resistance, which paved the way for type 2 diabetes through a series of metabolic changes. When stressed, her body assumed that more fuel was needed, and to accommodate, genes activated in cells to increase glucose uptake. During and following the assault, her body released surges of norepinephrine, adrenaline, and cortisol, which also increased blood glucose. The increases in cortisol triggered the breakdown of protein and its conversion in her liver to glucose. The excessive cortisol resulted in too much glucose floating around, which increased risk of insulin resistance. It was in the year prior to visiting her primary care physician that she developed type 2 diabetes.

We see the same dysregulating spiral with illnesses like arthritis, chronic pain, fibromyalgia, and chronic fatigue syndrome. As people become more depressed, their physical illness increases, which leads to greater depression. Decreased stress tolerance, increased anxiety, and depression associated with these chronic diseases combine to potentiate all of them together.

It is difficult to calculate how many people are depressed, suffer from cognitive deficits, and do not know that they are afflicted with spiraling dysregulations of their immune system, but it could be in the many millions. PICs tend to rise when physical health is compromised by autoimmune disorders, excess weight, and increases in response to stress. Stress and poor health impact multiple systems to create a spiral of decompensation within all levels of the mind-brain-gene feedback loops. The nonlinear interactions of all these factors put the person at greater risk for more serious physical and psychological disorders.

Psychotherapy in the twenty-first century must address the pandemic of acquired inflammatory diseases that have a devastating effect on mood and cognition. Lifestyle changes such as diet, exercise, and sleep (all addressed in Chapter 5) can affect the immune system. But before turning to those self-maintenance factors, we need to better understand a major part of the immune system, the gut.

GUT FEELINGS, THE ENTERIC NERVOUS SYSTEM, AND IMMUNE INTERACTIONS

Sylvia ambivalently sought psychotherapy at the suggestion of her primary care physician because of her vague complaints about intestinal discomfort and periodic sinus infections. He prescribed an ongoing "maintenance dose" of antibiotics to keep the sinus infection at bay. Willing to try anything to get over these physical problems, nevertheless at her first session she was more like a window shopper than someone seriously interested in psychotherapy.

Despite her nagging health problems and the time invested in trying to get well, she was devoted to her new business, a not yet financially successful bakery café. She ate breakfast there every morning, consisting of many new recipes for croissants and a "tall skinny latte." Her lunch included other "yummy" pastries. Because of the constant problems with her gastrointestinal tract, she consumed herbs that were advertised as being "gentle for the stomach." She also wondered if the fatigue, mild depression, and stress she experienced were related to "gut feelings" that her business was going to fail. Though her central concern revolved around her business, how was her gut communicating with her brain?

The gastrointestinal system, or gut, has been referred to as the "second brain" in the media and as the enteric nervous system (ENS) in the scientific literature. It includes a large group of neurons wrapped around the walls of the gastrointestinal system, which extends down from the esophagus to the distant colon, comprising approximately 500 hundred million nerve cells. The ENS maintains connections with the parasympathetic and sympathetic branches of the autonomic nervous system. It receives signals to stimulate digestive activity from the parasympathetic branch through the vagus nerve and to inhibit activity from the sympathetic branch. In other words, when a person is in fight-or-flight mode, the sympathetic branch acts to suspend digestive activity so that the person can devote all available energy to the challenges ahead. On the other hand, when stress subsides, the parasympathetic system promotes the rest-and-digest response.

Roughly 90 percent of the signals conveyed through the vagus nerve travel from the gut to the brain, while only 10 percent travel in the opposite direction. In other words, signals coming from the gut to the brain are more prevalent than those coming from the brain to the gut. The ENS is endowed with so-called dendritic cells that have tentacles that extend into the gut's interior. They can communicate with gut microbes that live near the gut wall. If they detect the presence of potentially dangerous bacteria, they can trigger a cascade of inflammatory reactions in the gut wall in an effort to control the pathogens, as in food poisoning.

Considering that our ancestors on occasion unknowingly ate contaminated food, a responsive immune system in the gut evolved to protect them from infection, illness, or even death. The connectivity between the gut and the brain became fine-tuned and responsive so that our ancestors would know that something was wrong. From this perspective, it is understandable that the gut makes up the largest component of the body's immune system: about 60 percent of the immune system is located around the intestines. with more dedicated cells than the thyroid, pituitary, gonads, adrenals, and circulating blood combined.

Many of the immune cells living in our gut are located in clusters in the small intestines in the Peyer's patches, as well as scattered throughout the walls of both the small and large intestines. A major part of the gut's immune system is located in a pouch-like area called the cecum and, dangling from it, the appendix. Overall, the cecum, appendix, and Peyer's patches represent a significant amount of immune activity.

The gut's abundant immune cells play a major role in signaling between the ENS nerves and the brain through the vagus nerve. These ongoing reports about the state of the gut and body in general comprise important interoceptive information. When inflammation occurs in the gut, many of the sensors become more sensitive to normal stimuli, especially in response to stress. As inflammation occurred in Sylvia's gut, her interoceptive information told her something was wrong, but not what. She interpreted these gut feelings as intuition that her business was in jeopardy. One of her responses was to consume excessive amounts of simple carbohydrates, through which she found temporary relief. Unfortunately, that temporary relief led to a long-term, out-of-control cycle of chronic inflammation. Over a lifetime, Sylvia's ENS adapted to stress that she encountered and to her diet by adding chemical tags to stress-response genes. As an adult she may have tended to sensitize gut reactions to stress, often complaining of "nervous gut feelings."

The increased production and release of pro-inflammatory cytokines (PICs) can contribute to a wide range of gut illnesses, including the aptly named inflammatory bowel disease. PICs travel to the brain either through binding to receptors on the vagus or through spilling into the bloodstream and traveling up to the brain like a hormone, traversing the blood-brain barrier and activating the microglia in the brain, which respond to and release more PICs.

> "Gut feelings," or interceptive information, may be misinterpreted as "intuition." Because so many factors contribute to inflammation in the gut, including poor diet, stress, and excessive antibiotic use, normalizing gut health optimizes mental health.

The Colony Within

The gut contains more than 100 trillion microorganisms, including bacteria, fungi, and archaea. If gathered together, they would weigh between two and six pounds, close to the weight of the brain. Outnumbering people on the planet by 100,000 times, the microbes in one person's gut perform many life-sustaining functions. The 1,000 bacterial species that make up what is called the gut microbiota contain 7 million genes, which means there are 360 bacterial genes for every human gene. Only a small fraction of our body's genetic content is of human origin, the reason many refer to us as superorganisms.

Approximately 90 percent of the 1,000 species of bacteria in the colon fall in either of two broad categories, Firmicutes or Bacteroidetes. Whereas the Firmicutes family increases fat absorption, because they are quite efficient at extracting calories from carbohydrates, the Bacteroidetes do not depend on carbohydrates and are more dominant in lean people. The ratio between Firmicutes and Bacteroidetes in the general public in the Western world has shifted in the direction of Firmicutes, turning on genes that add fat cells, increasing the risk for obesity, diabetes, and cardiovascular disease. Because people who are overweight tend to have a significantly higher ratio of Firmicutes to Bacteroidetes, they also tend to have abnormally high levels of leptin, because their extra fat cells produce more than lean people. Whereas with a lean person leptin signals the brain to decrease appetite, an obese person becomes resistant to its effects, and this critical feedback mechanism breaks down.

These microbes are so important that in 2008 the US National Institutes of Health launched an extension of the Human Genome Project to investigate the microbiome. The project has identified the complex interrelationships between a person's microbiota and overall health. It is now clear that the microbes in our gut participate in a wide variety of roles in our body, including the immune system, inflammation, detoxification, neurotransmitter signaling, vitamin production, and nutrient absorption. Together, the microbes, peptides, cytokines, and neurotransmitters interact with the brain. Approximately twenty different types of hormones are available to be released into the bloodstream at a moment's notice if needed. Neurotransmitters are synthesized in the gut, and up to 95 percent of the body's serotonin is stored there.

To understand why we must take gut bacteria into consideration for mental and physical health, it is important to keep in mind that gut microbiota differ widely among people, even identical twins. And within one individual, the diversity and abundance of microbes also vary over a lifetime. The diversity of microbes is low during the first three years of life while a stable microbiome is being established. Diversity reaches its peak during adulthood and wanes as we enter our senior years. Low diversity during the developmental years coincides with the vulnerability to neurodevelopmental disorders such as autism, while low diversity later in life coincides with the development of neurodegenerative disorders such as Parkinson's and Alzheimer's diseases.

Dysbiosis

Gut microbes constantly adapt to shifts in digestive fluids, acidity, available nutrients, and time intervals between meals. As the owner and head baker, Sylvia experienced stress that slowed down digestion. Microbes sensed changes and so activated genes that helped adapt to changing conditions. When her gut-microbiota-brain axis became unbalanced, she entered a state called dysbiosis, a microbial imbalance.

In a treatment meeting that included her primary care physician, we focused on the role of antibiotics in her dysbiosis. Since World War II, antibiotics have saved millions of lives. However, the more recent overuse of antibiotics has contributed to dysbiosis, and Sylvia's prescription of antibiotics for her sinus infection resulted in dysbiosis for her. This condition has been associated with a wide range of health and mental health disorders. We decided to phase out her prescription.

For healthy people, the lining of the gut is tight, so that microbes and other substances are prevented from passing into the abdominal cavity. However, the thickness and integrity of the thin mucus lining of the gut can vary depending on genetics, stress, and diet. Permeability of the lining, often referred to as "leakiness," can increase through stress and through excessive consumption of simple carbohydrates, saturated fat, and various food additives. When the combination of these factors occurred with Sylvia, her gut became leaky, and the inflammatory process spread throughout her body. This condition, referred to as metabolic toxemia, reduced her energy level while increasing her fatigue, pain sensitivity, anxiety, and depression.

A molecule referred to as lipopolysaccharide (LPS) is localized in the outer layer of gram-negative bacteria, which include Firmicutes and Proteobacteria. LPS levels increase in response to a high-animal-fat diet, and it is considered an endotoxin because it is a toxin that comes with bacteria and triggers a strong immune response. The resulting inflammation leads to a chain of events that increases the leakiness of the gut by allowing LPS to sneak between the cells in the lining of the gut and into the blood supply, where it interferes with the hormone insulin and promotes type 2 diabetes and heart disease. A regular diet of simple carbohydrates and saturated fat results in high levels of LPS, which stimulates cells that release PICs to make the gut even leakier, further spreading inflammation throughout the body, including the brain. The glial cells in the brain respond to the PICs by producing yet more inflammation, resulting in impaired cognition and dysphoric moods. Meanwhile, the hypothalamus becomes less responsive to satiety signals in the form of leptin from the gut, promoting overeating and further exacerbating the gut-brain system.

Contributors of Leaky Gut
A diet high in simple carbohydrates and saturated fats feeds the Firmicutes and lipopolysaccharides, leading to gut permeability.

Of significance to mental health, LPS is involved in a chain of events that increase depression (Maes et al., 2008). Because LPS makes the gut more permeable, PICs are more likely to gain access to and stimulate more PICs in the brain, increasing the risk for neurodegenerative disorders such as Alzheimer's disorder, lupus, multiple sclerosis, and autism, as well as depression. LPS has been shown to decrease the production of BDNF (Guan & Fang, 2006). This means that the brain-enhancing process of neurogenesis is blocked from making new cells in the hippocampus, and the person becomes prone to memory deficits and neurocognitive impairments. It is not surprising that up to three times as much LPS has been found in the plasma of Alzheimer's disease patients as in healthy adults (Zhang et al., 2009). Obese people tend to have much higher levels of LPS than do lean people, and it appears to trigger inflammation in their fat cells. Obesity is not simply layer upon layer of fat cells but fat tissue that has malfunctioned and become inflamed.

Lean people tend to have a high level of a type of gut bacteria referred to as *Akkermansia*. Though at best only representing 3–5 percent of gut bacteria, *Akkermansia* plays a significant role in supporting gut lining and regulating mucus production. The health of the mucus layer is critical for keeping other bacteria a safe distance from our gut's epithelial and immune cells. Despite the relatively low numbers of *Akkermansia* compared to other gut bacteria, if they are damaged or reduced in number the critical function they play in gut mucus lining maintenance is lost, resulting in inflammation. Not surprisingly, the lower the *Akkermansia* level, the higher the BMI. Another way to look at the effect of the relationship between the bacterium and the gut lining is that a higher level of *Akkermansia* acts to turn on genes to make more mucus, helping promote a healthy gut lining and preventing a leaky gut. A diet higher in saturated fat lowers *Akkermansia*, while a high-fiber diet promotes it. *Akkermansia* seems to thrive with the consumption of a type of fiber known as oligofructose, which is especially found in foods such as onions, bananas, leeks, and asparagus.

Mind, Brain, and Gut

Michael, introduced at the beginning of this chapter, felt that he had a "nervous stomach." The gut communicates with the brain through multiple channels. Microbes coat the razor-thin layer of mucus and cells of the inner lining of our intestines. This puts them in very close proximity to the gut's immune cells and cellular sensors that encode our gut sensations. Many of the brain-gut interactions occur here, ensuring that signals generated by the microbes are received by the brain and, in turn, those sent back to the gut are influenced by our emotions. Gut feelings are bidirectional, linking cells in the brain with those in the gut.

Interactions among the mind, brain, and gut can occur through top-down or bottom-up pathways that are emotionally laden via both branches of the autonomic nervous system. However, Michael primed his sympathetic branch and his neuroendocrine system so that he responded to assumed threats that were benign. For example, stress primed his hypothalamus to release corticotropin-releasing hormone and dampen the activity of gamma-aminobutyric acid, the principle inhibiting neurotransmitter in the body. This means that his stress tolerance and his ability to self-sooth tended to be low. In addition to the heightened tendency to easily trigger the release of stress hormones, such as norepinephrine and cortisol, which increased his hypervigilance and anxiety, corticotropin-releasing hormone stimulated his pituitary to release adrenocorticotropic hormone, which traveled through his bloodstream to the adrenal glands, which in turn released adrenalin, more norepinephrine, and cortisol. Simultaneously, the stress induced a gut reaction that impacted the composition and activity of the gut microbiota. He became prone to a wide range of sensations, including belly pain and gut contractions that resulted in frequent diarrhea. His stomach slowed down and even reversed itself, so he often felt like throwing up. Not only does the elevation in cortisol contribute to changes in the mix of gut bacteria, but it also increases the gut's permeability and triggers the release of PICs, which further increases the gut's permeability (Vanuytsel et al., 2014).

Early stress is also associated with alterations in microbiota and their metabolites, as well as with the stress circuits in the brain (Bercik et al., 2011). The neurodevelopmental changes resulting from early stress occur through multiple pathways, including epigenetic modifications of the brain-gut axis and stress-induced changes in gut microbiota and their products, which can further impair the brain. These abnormalities can begin to develop in utero and soon after birth. Interference in the infant's gut microbiome and a variety of challenges, including stress, nonvaginal delivery, unnecessary use of antibiotics, and unhealthy dietary habits during pre- and postnatal periods, can lay the groundwork for brain-gut disorders (Mayer, 2016). The pathophysiological syndromes that result from early stress and deprivation include the elevation of pro-inflammatory processes, as measured by C-reactive protein levels, decades later in life.

Throughout life chronic stress can stimulate the growth of many pathogens in a person's gut, making those pathogens more aggressive. Stress signals can also reduce the thickness of the mucus layer lining the colon, making it leakier, allowing microbes greater access to the gut's immune system, circumventing many of the gut's defensive mechanisms. Norepinephrine can stimulate the growth of bacterial pathogens that can cause serious stomach ulcers, gut infections, and sepsis. It can also activate genes in the pathogens that make them more aggressive and increase their odds of survival in the intestines. Certain gut microbes can modify norepinephrine into a more powerful form, intensifying its effect on other microbes (Mayer, 2016). Overall the combination of chronic stress, poor diet, and dysbiosis can increase the leakiness of the gut leading to greater metabolic toxemia.

In the epigraph to this chapter, Taisen Deshimaru's quote, "Think with your whole body," referred to being consciously present with every ounce of your being. Based on the concepts described in this chapter, we actually do think and feel with our whole body. The interactions between the immune system, including within our gut, and the brain are dynamic. Dysregulation within the gene-immune-brain-mind feedback loops can have confusing and devastating effects on mood and cognition.

Psychotherapy by necessity must focus on the behavioral health of clients. This means that self-maintenance factors such as diet, exercise, and sleep, the focus of Chapter 5, need to be addressed as foundational factors to mental health.

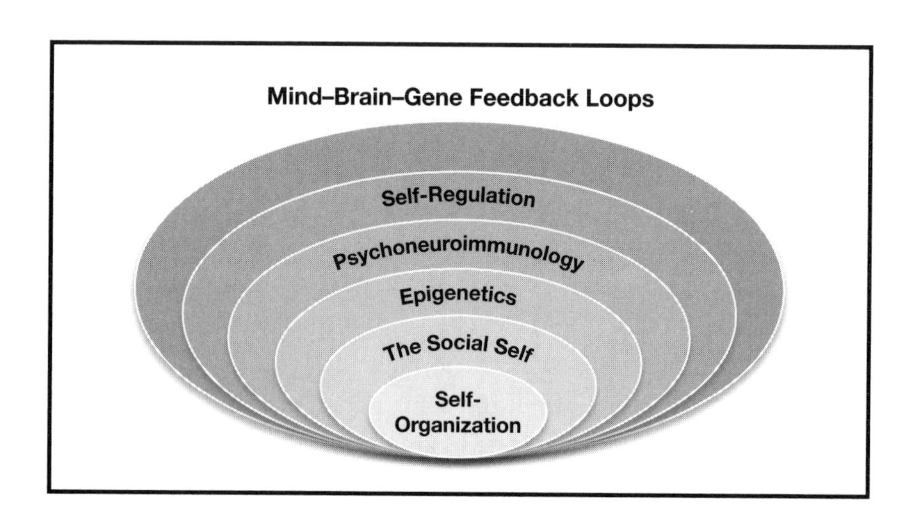
Mind–Brain–Gene Feedback Loops
Self-Regulation
Psychoneuroimmunology
Epigenetics
The Social Self
Self-
Organization

FIVE

Self-Maintenance

> And God said, "Behold, I have given you every plant yielding seed that is on the face of all the earth, and every tree with seed in its fruit. You shall have them for food."
>
> —Genesis 1:29

Self-maintenance systems play a major role in gene expression and immune system functioning. A nutritionally impoverished diet, lack of exercise, and poor-quality sleep are associated with unhealthy gene expression and shortened telomeres, all undermining mental health. This combination of factors represents the foundation of health. Without addressing these factors immediately, the traditional psychotherapy interventions fail to gain traction.

Like first responders, the self-maintenance systems must be addressed right away to stabilize and ensure that no more damage is done—for this reason this chapter is up front rather than tucked away in an appendix. In fact, people who work as first responders cannot respond effectively if these self-maintenance factors are undermined.

For example, Mike, who worked as a police officer, came in to see me because of his irritability, decreased stress tolerance, and restless sleep. "I'm just not myself," Mike told me. "I've turned into one of those grumpy old cops you see on TV." Serving in a high-crime neighborhood, Mike found his job extremely stressful and exhausting. He often joked that that while his trips to the donut shop were a cliché, he needed the "extra boost" the sugar gave him but complained about "crashing" two hours later. Mike's wife, Molly, worked as a dispatcher for the police department and was well aware of the gang violence and drive-by shootings on his beat. They both agreed that he should work his current beat for one more year in order to earn promotion to detective. In the meantime, it was essential that Mike learn to take better care of himself.

Molly told me during a phone call that Mike started the day with an energy bar and a cup of coffee with three spoonfuls of sugar. She described his typical lunch as a few fried egg rolls or chicken nuggets. He insisted on bringing home fish and chips, fried chicken, and cake.

During our third session, when we reviewed his diet, Mike told me that what he ate had nothing to do with his grumpiness. He stared at me and said, "I hope you're not a health nut like my wife!" When I told him that I hoped to help him deal with his stress at work, he seemed to slightly lower his guard. As a sarcastic joke, he said that he would shift from donuts to blueberry muffins at the donut shop, thinking that they were healthier. With a playful smile I pointed out that the muffins were loaded not only with empty calories but also trans-fatty acids. He replied, "I thought you food fanatics say that blueberries are good for you!" We both had a good laugh.

THE PERILS OF THE MAINSTREAM DIET

Mike's stress level was higher than that of police officers in small towns with less violence. His resistance slowly dissolved by my asking if his poor diet was undermining his stress tolerance. The short answer was that Mike's body was not prepared for the constant onslaught of stress that triggered his need for cheap energy.

It is important to keep in mind that our bodies did not evolve to consume processed foods. Prior to eleven thousand years ago, our ancestors lived a hunter-gatherer subsistence lifestyle, with a diet consisting of plants, nuts, berries, fish, and lean game. While our bodies have not changed since the Paleolithic era, our diets have, with negative results. The Western diet—Mike's diet—consists of deep fried foods, refined carbohydrates, and processed foods containing sugar and high-fructose corn syrup, which spike glucose levels. When the body digests carbohydrates, it produces glucose, which accumulates in the bloodstream. In response, the pancreas releases insulin, which signals the liver and muscles to convert glucose into a starch-like molecule called glycogen. We also possess an enzyme called uricase that stores up energy is fat cells. This efficient method of energy storage worked quite well with a Paleolithic diet and a small supply of food. Mike, like too many people, was excessively "storing energy" in the form of extra fat cells. In his case, he had an extralarge belly, leaching out pro-inflammatory cytokines.

In addition to consuming fast food with empty calories, he drank one, sometimes two Big Gulp sodas. A couple of hours after Mike's brain was assaulted with surges of glucose from the simple carbohydrates he consumed, his blood sugar and energy level took a nose dive. And so was there a loss of fuel for his prefrontal cortex—the CEO of his brain. With less energy for his executive network, his default-mode network replayed all the mishaps of the day. Nervously focused on the past instead of the present and anticipating what he might encounter in the immediate future made him prone to make mistakes, sometimes dangerous ones.

When blood sugar (glucose) drops below 50 milligrams per milliliter, the following symptoms may occur:

- Free-floating anxiety
- Mood instability
- Lethargy
- Shakiness
- Lightheadedness
- Irritability
- Rapid heartbeat
- Difficulty concentrating
- Memory problems

Fructose, unlike other sugars, produces uric acid when it is broken down in cells and blunts the effect of leptin, the hormone that tells the brain to stop eating. Elevated fructose essentially flips a switch in the body, causing it to hoard fat and raising blood sugar and blood pressure. A diet high in refined, processed carbohydrates and fats is associated with high levels of C-reactive protein, a common measure of inflammation (Liu et al., 2002).

Many processed foods and many "fruit" drinks contain fructose. Fructose has been shown to increase lipopolysaccharide by as much as 40 percent (Benros et al., 2013). As noted in Chapter 4, the combination of lipopolysaccharide and the gut bacteria Firmicutes, which feed on simple carbohydrates, increases gut permeability—leaky gut contributes to chronic inflammation.

Soon after our first session, Mike was involved a shooting during a robbery and was reassigned for six months to a desk job. Thinking he no longer needed help with stress, he did not return until six months later. In the meantime his diet, high in simple carbohydrates, combined with a sedentary lifestyle contributed to insulin resistance and obesity, which led to a host of health problems affecting his mental health, including metabolic syndrome and type 2 diabetes. The associated increases in blood pressure, cholesterol, and triglycerides elevated his risk for developing heart disease. In addition, a rise in uric acid increased his blood pressure by oxidative stress, which constricted his blood vessels, forcing his heart to pump harder. Also, the increased uric acid caused low-grade injury and inflammation in his kidneys, which made them less able to excrete salt and increased his blood pressure even more.

Glucose must be balanced so that the pancreas, liver, thyroid, adrenals, pituitary, and brain function properly. Keeping glucose levels in balance is critical so that the pancreas can deliver the right amount of insulin to regulate fuel delivery to cells. If glucose is elevated too long, a reaction called glycation occurs that causes sugars to attack proteins, causing inflammation in cell membranes and impairing interactions between neurons (Epstein, 1996). Advanced glycation end products (AGEs) additionally shorten the life span of cells by causing inflammation and creating free radicals, which result in cognitive and emotional deficits. AGEs also lead to structural damage to mitochondria, the energy factories within cells, which produce cellular energy in the form of ATP (Kikuchi et al., 2003).

Mike not only complained about fatigue and depression but was also referred to a neuropsychologist because of persisting and increasing memory problems. He thought he was developing Alzheimer's disease but was actually elated when the neuropsychologist described the detrimental effects of his diet and his sedentary job. It was at that point that he returned to me to address his self-maintenance.

Glycemic Load

A useful measure of glucose level, glycemic load (GL), relates to the expected rise in blood sugar caused by a particular food. Foods with a high GL, such as white bread, instant oatmeal, French fries, and fruit juices, lead to greater inflammation, which can have destructive effects on the brain and impact cognition, memory, and mood. Consistent with these findings, a diet high in simple carbohydrates is associated with increased swelling and decreased brain-derived neurotrophic factor (BDNF) in the hippocampus, which is central to explicit memory. This is one of the reasons that people with diabetes are at increased risk for developing memory problems.

In the last forty years the standard Western diet has been marked by an exponential increase in soft drinks and fruit drinks. They contain sugar or high-fructose corn syrup, which trigger free radical products of damaged fatty acids, called isoprostanes, which rise 34 percent just ninety minutes after consumption (Sampson et al., 2002). When fatty acids such as arachidonic acid are damaged, isoprostanes appear in tissues and cause free radical activity, which chips away at the cell membrane. To put this in perspective, levels of isoprostanes are nine times higher one day after a brain injury than before the injury (Pratico et al., 2002). High levels of isoprostanes are found in the cerebral spinal fluid, plasma, and urine of people with Alzheimer's disease, which implies that they are a possible predictor of the disease.

An increase in glycemic load and a blood marker of oxidative stress, malondialdehyde, have been shown to damage essential fatty acids (Hu et al., 2006). Because essential fatty acids are critical to the structure and health of our brain, this combination of inflammatory factors and oxidative stress is corrosive to the brain and thus to mood and cognition.

Essential Fats

After helping Mike stabilize his diet, I was invited to give a presentation on brain health and stress tolerance to the entire police department. Just to get the discussion started, I told the officers that the next time someone called them fat heads, say thanks! A large proportion of the brain is composed of fats, so despite the food fads, fat consumption is critical to a healthy diet. However, during the past century the amount of healthy fats in the form of essential fatty acids in the Western diet has dramatically declined. In the place of good fats we are now consuming destructive fats, including animal fats, trans-fatty acids, vegetable oils, and processed foods.

A diet high in saturated fat has increasingly been associated with cardiovascular disease and dementia, and the World Health Organization has reported its association with cancer. When combined with social stress, a diet high in saturated fat leads to deterioration of the hippocampus associated with the retraction of dendrites (Baran et al., 2015).

We all had a good laugh when caterers began to set up. The lunch buffet featured hamburgers, French fries, onion rings, and fried fish patties. One of the lieutenants said that the menu was not an anomaly. Like Mike, many of the officers admitted to consuming fried foods at home on a regular basis, not knowing of the multiple adverse consequences. Trans-fatty acids are formed when an unsaturated vegetable fat is heated for long periods of time in a metal container, such as deep frying. Trans-fats, which are structured differently than essential fatty acids, tend to be solid at body temperature and act like saturated fat, all of which are corrosive to cell membranes. Consuming trans-fatty acids decreases the levels of healthy fats, so the uptake of trans-fatty acids into the brain doubles.

The Destructive Effects of Trans-Fatty Acids

- Are absorbed directly by the nerve membranes, making cells more rigid and inflexible
- Block the body's ability to make its own essential fatty acids
- Alter the synthesis of neurotransmitters such as dopamine
- Negatively affect the brain's blood supply
- Increase bad (LDL) cholesterol while decreasing good (HDL) cholesterol
- Increase plaque in the blood vessels
- Increase blood clots
- Increase triglycerides, which cause the blood to be sluggish and reduce the amount of oxygen to the brain
- Cause excess body fat, which can have a destructive effect on the brain through chronic inflammation and pro-inflammatory cytokines

Despite our good laugh together over the menu, most of the officers went up for seconds. If we eat more calories than we burn, our triglycerides levels rise. Triglycerides are a type of fat found in the blood. The body converts calories the body does not need right away into triglycerides, which it stores in fat cells. Between meals, hormones release triglycerides for energy. Elevated triglycerides are associated with multiple health problems, including depression, while lowering the triglyceride level is associated with the alleviation of depression.

When essential fatty acids are not balanced, pro-inflammatory cytokines become overactivated, turning the immune system against cells, such as occurs with autoimmune disorders. As noted in Chapter 4, increases in pro-inflammatory cytokines are associated with cognitive problems, anxiety, and depression.

Omega-3 fish oil supplementation is associated with decreased pro-inflammatory cytokines and free radical damage. Some of the anti-inflammatory effects of omega-3s appear to be related to altered expression of genes encoding inflammatory mediators, as well as lower oxidative stress.

Omega-3 Essential Fatty Acids and Functions in the Brain

Eicosapentaenoic acid (EPA)

- Affects receptor functioning
- Is used to manufacture more synapses
- Is involved in the conversion of L-tryptophan to serotonin and in the control of its breakdown
- Reduces the risk of blood clots and the amount of arterial plaque
- Has anti-inflammatory effects
- Helps lower triglycerides (fats in the blood)
- Lowers blood pressure

Docosahexaenoic acid

- Is present in high concentrations in synaptic membranes and mitochondria
- Is essential for synaptic transmission and membrane fluidity
- Is critical in keeping cell membranes soft and flexible
- Helps hold receptors in place—soft and flexible membranes can alter the shapes of the receptors, which is critical for neuroplasticity

Omega-3s have been associated with promoting BDNF, which as noted in Chapter 4 plays a critical role in neurogenesis and neuroplasticity, serves as a neuroprotective agent, and is critical for memory and new learning. While inflammation and oxidative stress interfere with BDNF production, omega-3s can lower both inflammation and oxidative stress. Low levels of BDNF are associated with neurological damage and depression, and the omega-3 EPA can help maintain BDNF. On the other hand, the combination of high cortisol, disrupted gut microbiota, high levels of inflammation in the gut, systemic inflammation, and lower levels of BDNF has been consistently associated with anxiety and depression (Glaus et al., 2014).

Excessive doses of omega-3 are counterproductive—lower doses are more effective than higher doses. Consumption of over 2 grams per day is related to *higher* levels of depression (Pect & Stokes, 2005).

Essential fatty acids can also help the brain facilitate the so-called second-messenger system that activates when neurotransmitters penetrate the fatty membrane of the cell and trigger secondary emissaries that reach into the nucleus of the cell, where they turn genes off or on, to send chemicals back outside the cell and create yet even more reactions.

The Essential Neurochemical Cornucopia

Brains capable of generating balanced moods and clear thinking need a diet containing essential amino acids, vitamins, and minerals to produce the cornucopia of brain chemistry optimal for the brain. Because neurotransmitters are synthesized from precursor amino acids that serve as their building blocks, a diet lacking these critical amino acids diminishes neurochemistry, contributing to deficits in memory, attention, and mood. Specific amino acids serve as the building blocks for neurotransmitters.

Amino Acid Precursors to Neurotransmitters

Amino Acid	Neurotransmitter	Effects
L-Trytophan	Serotonin	Improves sleep, calmness, and mood
L-Glutamine	GABA	Decreases tension and irritability
L-Phenylalanine	Dopamine	Increases feelings of anticipation of pleasure and drives motivation
L-Phenylalanine	Noreprinephrine	Increases energy, feelings of pleasure, and memory

Given the huge amount of information available about foods, and the food fads that have come and gone, what can we consider the most consistently supported recommendations? A diet with low levels of saturated fat and high in fruits and vegetables, such as the Mediterranean diet, has been associated with a lower risk of age-related cognitive impairment (Feart et al., 2009). When comparing the Mediterranean diet alone or in combination with regular exercise, the addition of exercise lowers the risk of developing dementias such as Alzheimer's disease (Scarmeas et al., 2009). People who adhere to a Mediterranean diet, consisting of fresh fruits and vegetables, anti-inflammatory fats, and proteins have been found to have a lower rate of depression (Sanchez-Villegas et al., 2009).

Adherence to the Mediterranean diet has been associated with telomere maintenance and associated health status and longevity (Boccardi et al., 2013). The Asian equivalent, sometimes called the Okinawan diet, which consists of high consumption of fish, seaweed, and vegetables and low consumption of simple carbohydrates and red meats, is also associated with telomere maintenance (Lee et al., 2015). A "plants plus" diet oriented around vegetables with only "sensible carbohydrates" and minimal saturated fats is best for mental health (Borysenko, 2014).

Inflammation and Diet

Diet plays a major role in stress tolerance, which depends on a healthy microbiome. The serotonin-containing cells in the enteric nervous system play a role in linking the food we consume and activity in our gut to our brain. The microbes that feed on the food influence our mood through gut-brain interactions. For example, a diet low in tryptophan, the amino acid precursor to serotonin, results in less serotonin and potentially depressed mood. The gut bacterium *Bifidobacterium infantis* plays a role of making tryptophan available to be synthesized into serotonin. Microbial metabolites can increase the production of serotonin in the enterochromaffin cells in the gut. The vagus nerve signals the brain when enough is available, having an effect on mood and stress tolerance.

Though so-called comfort foods do provide momentary comfort during periods of stress, a chronically stressed person experiences only a temporary inhibition of the stress response system. The benefits are short-lived and are followed by many more hours of emotional discomfort, as well as cognitive fog, the growth of fats cells, and even obesity (Tomiyama et al., 2011).

Various gut hormones act through feedback loops to the reward pathways in the brain that boost appetite. Those associated with overeating include dopamine-containing cells, which increase appetite, while certain appetite-suppressing signals decrease dopamine release. These dopamine pathways follow the main route through which all addictive behaviors activate: the nucleus accumbens. The reward system and the appetite cycle worked well together for hunter-gatherers to promote motivation for food seeking. However, with constant food availability, many people find it difficult to resist the temptation to eat out of habit and addiction. If the food is salty, fatty, and/or sugary, it becomes even a greater source of addiction. Fast food outlets build their menus to incorporate these factors and yield great profits. Consistent with our vulnerability to fast foods, many processed food companies have developed recipes that we "cannot eat just one." The consequences of regular consumption of such empty-calorie foods include dysbiosis of the microbiome, low-grade inflammation (metabolic endotoxemia), autoimmune disorders, and anxiety and depression.

Shifting the microbiome to promote mental health is consistent with the benefits of the Mediterranean or Okinawan diet. For example, healthy microbes in the colon metabolize many of the undigested plant-derived carbohydrates into short-chain fatty acids. Among them is butyrate, so named because of its buttery odor. It plays a key role in providing nutrients for the cells lining the colon and protects the brain against the destructive effects of low-grade inflammation caused by a high-fat diet and artificial sweeteners (Mayer, 2016). Butyrate aids in plugging the leaks in the gut by acting as a messenger to turn up the volume on genes to make more protein chains in the gut wall to tighten it. Butyrate causes the release of leptin for satiety, which plays an important part in preventing obesity.

MOVEMENT, OUR EVOLUTIONARY IMPERATIVE

Prior to 11,000 years ago, our ancestors walked approximately 10 miles per day in search of food. Because the portion of the human genome that determines our basic physiology has remained unchanged, it leaves us poorly adapted to our current sedentary lifestyle and diet (Cordain,

Gotshall, & Eaton, 1998). Many of the genes selected for the survival of our ancestors now impair our health. The result is that physical inactivity makes us more susceptible to chronic illnesses. This largely accounts for the increase in the prevalence of such illnesses as atherosclerosis, certain cancers, obesity, and type 2 diabetes (Booth & Neufer, 2005). During my presentation to the police officers I was informed that the department gave them free membership to local gyms. By a show of hands, only two indicated they took advantage of this perk. Not coincidently, those two looked the healthiest, with bright eyes and taking frequent notes during the seminar.

Well over 60 percent of Americans are overweight, and only 25 percent are physically active. More troubling is that only 10 percent are active enough to achieve fitness. The critical point that must be addressed in psychotherapy is that exercise is a key component of improving mental health.

Regular exercise not only burns off excess calories and weight but also increases the insulin receptor sites throughout the body to aid in the metabolism of glucose. The initial burst of pro-inflammatory cytokines after exercise seems to trigger the body's anti-inflammatory mechanisms to keep inflammation in check (Goebel et al., 2000). As we age the potential for inflammation rises. However, with older adults, resistance training, flexibility training, and aerobic exercise all lead to improved levels of optimism and the lifting of depression, but only aerobic exercise led to lowering of inflammation as measured by reduced levels of C-reactive protein and interleukin-6.

Exercise offers one of the best antidepressant and antianxiety treatments. It costs nothing and offers only good "side effects." A wide variety of factors contribute to these positive effects. Exercise promotes increased efficiency of the neurotransmitters serotonin and dopamine, which are the very neurotransmitters targeted by the major antidepressant medications. Exercise upregulates these neurotransmitters, in contrast to medications, which downregulate them. In other words, exercise promotes their activity long afterward, promoting less relapse, in contrast to medications, which are notorious for relapse of depression after "treatment."

In addition, as described in Chapter 9, depression is associated with abnormal default-mode activity. Exercise has been shown to decrease and normalize default-mode network activity (McFadden et al., 2013). Meanwhile, exercise has been shown to bolster executive network activity across all age groups.

Exercise as a Treatment for Depression

- An Alameda County study of 8,023 people tracked them for twenty-six years and found that those who didn't exercise were 1.5 times more likely to be depressed.
- A Finnish study of 3,403 people found that those that exercised two to three times per week were less depressed, angry, stressed, and cynical.
- A Dutch study of 19,288 twins and their families showed that those that exercised were less anxious, depressed, and neurotic and more socially outgoing.
- A Columbia University study of 8,098 found the same inverse relationship between exercise and depression
- An Ohio State study found that 45 minutes of walking per day, 5 days per week (heart rate at 60–70 percent of maximum) lowered scores on the Beck Depression Inventory from 14.81 to 3.27, compared to no change for controls who were depressed nonwalkers.
- A University of Wisconsin study found that exercise in the form of jogging was as effective as psychotherapy for moderate depression. After one year 90 percent of the exercise group were no longer depressed but 50 percent of the psychotherapy group were depressed.
- A Duke University study found that exercise was as effective as Zoloft. At 6-month follow-up, exercise was 50 percent more effective in preventing relapse. Combining exercise and Zoloft added no benefit regarding relapse (Babyak et al., 2000)
- A National Institute of Mental Health panel concluded that long-term exercise reduces moderate depression.

Exercise Epigenetics

Because our genes evolved with the expectation of a certain threshold of physical activity for normal gene expression, regular exercise restores the homeostatic feedback mechanisms toward the normal physiological range of our ancestors. Daily exercise normalizes gene expression critical for not only general health but also mental health.

Movement alone is a step in the right direction (pun intended). The trend toward weakening of skeletal muscle has been estimated to range from 8.8 percent in people under seventy years of age to 17.5 percent in people older than eighty years. People with weak muscles have a greater risk of mortality and health problems. Whereas contracting muscles increases their glucose uptake, reduced physical activity is associated with a rapid development of insulin resistance.

> A large number of genes in the skeletal muscles are activated after exercise. Based on the duration of the exercise, the genes fall in three categories. "Stress-response genes" are activated during the later phases of exercise. Their protein concentrations rise quickly to high levels and return to normal levels very quickly after exercise. These acutely sensitive genes encode proteins that are part of the general response to stress in all types of cells and include heat-shock proteins and some transcription factors—immediate-early genes. The second category is referred to as "metabolic-priority genes" that are required as a consequence of metabolic stress, such as when muscle glycogen and blood glucose levels are low. These genes can also be expressed at high levels, usually peaking after a few hours and returning to normal levels after twenty-four hours. The third category is referred to as "metabolic/mitochondrial enzyme genes." These genes encode protein whose function is to convert food to energy. They produce much lower concentrations but do not return to normal levels for up to one week. These genes play a role in muscle plasticity, which involve the increase in mitochondrial and capillary concentration.

Exercise and Anxiety Reduction

There are multiple ways that exercise serves as an effective treatment for anxiety. At the most basic level, the buildup of stress increases muscle tension, whereas exercise relaxes the resting tension of muscle spindles, breaking the stress feedback loop to the brain. Exercise lowers blood pressure by increasing the efficiency of our cardiovascular system, pumping healthy levels of oxygen to the brain and increasing the health of the capillaries. (Swain et al., 2003). As our heart rate increases during exercise, so does the hormone atrial natriuretic peptide, which tempers our body's stress response by going through the blood-brain barrier and attaching to receptors in the hypothalamus to tone down the hypothalamic-pituitary-adrenal axis activity. It also acts on our amygdala and the locus ceruleus to dampen the effects of corticotropin-releasing hormone, part of the neuroendocrine stress system that can induce anxiety.

Exercise upregulates a variety of neurotransmitters and neurohormones to buffer stress and lower anxiety. During and after exercise there is a release of endorphins and neuropeptides that bind to opioid receptors in the brain, inducing the so-called runner's high, with potent analgesic effects and feelings of easiness. Serotonin levels increase when our body breaks down fatty acids to fuel our muscles.

> Exercise upregulates GABA. Simply moving the body triggers the release of GABA, whereas benzodiazepines downregulate the GABA system after targeting their receptors. They contribute to tolerance and withdrawal, the hallmarks of addiction. When they wear off the ill-informed patient experiences "anxiety sensitivity" and feels more anxiety during early withdrawal, sometimes prompting the mistaken assumption that the patient "needs" the drug. Benzodiazepines also contribute to depression. The bottom line is that taking these drugs is similar to wading into a warm pit of quicksand. He feels good initially, but soon he is sucked down into the abyss of depression and, ironically, more anxiety.

Brain Enhancement and Repair

Exercise promotes multiple brain enhancing mechanisms, which are critical for brain health. They include neurotrophic factors that rebuild brain structure and enhance blood flow. The two neurotrophic factors which have received the most attention are BDNF and glial-cell-derived neurotrophic factor. Both are involved in neuroplasticity and neurogenesis. BDNF has been shown to produce new neurons in the hippocampus and prefrontal cortex. Glial-cell-derived neurotrophic factor (GDNF) is involved in new glial cells. Both are released after exercise, and both enhance cognition and diminish depression.

Brain Health and Exercise

Mechanism	Impact
Gene expression	Neuroplasticity
Brain-derived neurotrophic factor (BDNF)	Neuroplasticity, neurogenesis (Adlard, Perreau, & Cotman, 2005)
Insulin-like growth factor-1 (IGF-1)	Neural energy utilization (Carro, Trejo, Busiguina, & Torres-Aleman, 2001)
Nerve growth factor (NGF)	Enhanced neuroplasticity (Neeper, Gomez-Pinilla, Choi, & Cotman, 1996)
Vascular endothelial growth factor (VEGF)	Strengthening blood vessels. Also, enhanced neurogenesis (Fabel et al., 2003)

These repair factors also help prevent the damaging effects of chronic psychological stress by keeping the stress hormone cortisol in check. High levels of cortisol have a corrosive effect on the hippocampus, resulting in its atrophy. Because the hippocampus is critical for laying down new memories, chronically high stress can result in a memory deficit and the inability to turn off the hypothalamic-pituitary-adrenal axis.

The bottom line is that exercise is an indispensable adjunct to therapy. Clients who do not engage in regular exercise undermine their recovery. Exercise reduces stress, anxiety, and depression by pumping up the levels of neurotransmitters GABA, acetylcholine, serotonin, dopamine, and norepinephrine that keep us calm, positive, and energized. It also increases the repair factors to maintain and build a healthy brain.

THE NEED FOR QUALITY SLEEP

When I asked the police officers about their sleep, most reported that they had too little of it because of insomnia or frequent awakenings during the night. Sleep medication, either prescribed or over the counter, were common, and the officers reported diminishing benefits after a few days. Many of the officers reported waking up in the morning feeling groggy and then going directly to coffee or energy drinks.

When I informed them that there were a variety of sleep hygiene techniques that can help them get good-quality sleep. One said, "Crappy sleep is a hazard of the job." The rest nodded, punctuated with sarcastic laughs.

We spend one-third of our lives asleep. The quality of that sleep plays a major role in our cognitive and emotional experience during the other two-thirds of our lives. Failure to follow a healthy sleep schedule increases mortality risks, mood disorders, and even cancer prevalence (Evans & Davidson, 2013). Many companies and, surprisingly, health care programs have failed to heed the results of research on rotating shift work, which show not only major health risks but also a wide range of work performance problems: not just more accidents and health costs but also cognitive deficits such as poor and risky decisions (Venkatraman, Chuah, Huettel, & Chee, 2007).

Sleep provides a means through which the brain can clear out toxins. During sleep a plumbing system for cerebral spinal fluid called the glymphatic system opens up, letting fluid flow more efficiently through the brain. Glia cells control the flow through the glympathatic system by shrinking or swelling. Norepinephrine, an arousing neurotransmitter, controls cell volume: with a low level of norepinephrine, such as during healthy sleep, fluid flow increases because of the greater space between the cells. Because toxic molecules, including beta-amyloid involved in neurodegenerative disorders, tend to accumulate in the space between brain cells, a healthy glymphatic system provides a method to clear them out (Xie et al., 2013).

As an illustration of the importance of good-quality sleep, chronic insomnia is associated with shorter telomeres (Blackburn & Epel, 2017). In other words, sleep is critical for the overall maintenance of the brain, and without good-quality sleep we age quicker and suffer cognitive and mood regulatory deficits.

Functions of Sleep

- Clear out toxins
- Protein synthesis (Ding et al., 2004)
- Synthesis and transport of cholesterol (Cirelli, 2005)
- Expression of molecules associated with synaptic plasticity (Taishi et al., 2005)
- Increased long-term potentiation (Cirelli & Tononi, 2008)
- Expression of genes (Cirelli & Tononi, 2008)
- Memory consolidation

The Whitehall Study of thousands of British civil servants found that men who slept five or fewer hours most nights had shorter telomeres than those who slept more than seven hours. These results were independent of obesity, socioeconomic status, and depression (Jackowska et al., 2012). Studies suggest that an average of seven hours of good-quality sleep per evening is optimal (Ferrie, et al, 2011). Six hours of sleep is reported as the minimum biological requirement and for this reason is referred to as the "core sleep." However, people who sleep more than nine or less than six hours, while not constituting a sleep disorder, are at greater risk of cardiovascular disorders, hypertension, type 2 diabetes, and other disorders correlated with inflammation.

Sleep deprivation causes

- Impaired neurogenesis
- Restricted ability to clear glucose from blood up to 40 percent

- Heightened risk of type 2 diabetes
- Increased blood pressure
- Weight gain
- Depression
- Anxiety
- Shortened telomeres

Sleep deprivation results in a drop in the ability to clear glucose through insulin, decrease leptin (which would normally inhibit hunger), and increase ghrelin (which increases hunger). Taken as a whole, these factors are consistent with the overall concept that sleep deprivation contributes to weight gain, a fact now supported by at least fifty studies (Strickgold, 2015). Sleep deprivation, even after one week, can promote weight gain through an increase production of ghrelin, matched by a decline in the production of the hormone leptin. The increased appetite associated with sleep loss tends to promote the consumption of foods high in simple carbohydrates and calories, at the expense of hunger for fruits, vegetables, and high-protein foods. Sleep-deprived people may consume up to 33–45 percent more calories than non-sleep-deprived people.

Sleep deprivation destabilizes the functioning of mental operating networks, for example, with deficits in attention and task completion of the executive network. There appears to be abnormal "resting state" activity associated with the default-mode network (Gujar, Yoo, Hu, & Walker, 2010). Correspondingly, sleep deprivation reduces default-mode network (DNM) connectivity (De Havas, Parimal, Soon, & Chee, 2012). Sleep deprivation has been shown to alter the activation of the salience network, and its reward process is associated with overeating (Fang et al., 2015).

Given that when memory is reenacted it becomes unstable, the mood that we are in during sleep reconsolidates memories within the context of that mood. In fact, the concept of memory reconsolidation has itself changed to memory evolution. When sleep deprived, we form twice as many memories of negative events as positive events. Coupled with decreased executive network activity and increased DMN activity, this fuels the DMN tendency to ruminate on negative memories and increase depression. Sleep can result in memory evolution of traumatized feelings that increase with nightmares, such that acute stress disorder evolves into posttraumatic stress disorder (PTSD).

People who suffer from trauma generally experience fragmented sleep and less slow-wave (stage 4) sleep. Slow-wave sleep is critical for consolidating explicit memory, while REM sleep is critical for consolidating procedural memory. Sleep deprivation results in impairments in attention, new learning, and memory. During sleep unstable memory traces are reconfigured into more permanent ones for long-term storage (Frank, Issa, & Stryker, 2001).

In addition to insomnia, people who suffer from trauma wake up feeling sleep deprived as well as fatigued. They also have reduced time in REM sleep, in part because of the increase in norepinephrine. Because norepinephrine is generally inactive during REM for people without PTSD, the failure to shut down norepinephrine in people with PTSD leads to an incomplete entry to REM sleep. And as noted above, high levels of norepinephrine impair the glymphatic system's ability to clear out toxins.

One night of sleep deprivation can negatively shape emotional memories, impairing a person's ability to recognize positive or neutral words by 50 percent (Strickgold, 2015). In other words, when sleep deprived we are prone to lay down twice as many memories of negative events as positive ones, which contributes to depression.

Sleep loss, as well as sleep disturbances, tends to result in impairments in the immune system. It has long been known that people who experience less than 6 hours of sleep tend to be more susceptible to illness, such as viral infections. Sleep loss and disturbances have been associated with the reduction in lymphocyte responses and natural killer cell activity. Insomnia has been associated with elevations in inflammatory markers such as interleukin-6 and tumor necrosis factor alpha (Vgontzas, et al., 2002). Proinflammatory cytokines are associated with inducing sleep, particularly slow-wave deep sleep. Cortisol levels are the lowest during slow-wave sleep, and the levels rise late in the sleep cycle during REM periods.

This may be one of the reasons that not just daytime sleepiness and fatigue but also achiness and depressed mood follow a night with insomnia. Inflammatory biomarkers such as C-reactive protein and interleukin-6 are associated with metabolic disorders and have been shown to be associated with sleep dysregulation.

As noted throughout this book, for many people health problems that disrupt the length and quality of sleep tend to be associated with mental health problems. For example, people with sleep-disordered breathing tend to be overweight. Common and obstructive sleep apnea is often associated with fat buildup or loss of muscle tone. Sleep-impaired breathing represents a range of sleep disorders that tend to be underdiagnosed and strongly associated with cognitive decline and mood disorders. People with sleep apnea are especially vulnerable. They suffer from an increased risk of coronary heart disease, hypertension, type 2 diabetes, mental health disorders, cognitive impairment, and dementia. Likewise, all of these health problems, as well as sleep dysregulations, are associated with increased levels of inflammation. However, when people with sleep apnea are put on a continuous positive airway pressure (CPAP) device, cognitive decline and depression can be arrested (Osorio et at., 2015).

RESETTING THE CIRCADIAN RHYTHM

When not complicated by artificial light, our sleep pattern and circadian rhythm follow the daylight length. When full-spectrum light hits our retina, this signals our pineal gland that it is daytime. Our pineal gland responds by suppressing its release of melatonin, signaling that it is not time to sleep. Alternatively, when it is dark, our pineal gland releases melatonin to induce sedation.

> Resetting the circadian rhythm represents a key part of treating insomnia. Because the amount of light to which we are exposed during the daytime affects sleep, clients with insomnia should maximize exposure to bright light in the daytime to set the body clock to match the natural day/night cycle. Exposure to bright natural light in the late morning encourages lower body temperature in the middle hours of the sleep

cycle and promotes staying asleep. With early morning awakening, clients should expose themselves to bright light, helping ensure that the pineal gland does not produce melatonin throughout the day and that his body temperature will be the lowest during sleep. If clients suffer from insomnia they should not use a computer in the late evening because this is essentially looking at light, which tricks the brain into adjusting to a daytime pattern, suppressing the pineal gland's secretion of melatonin. Clients with insomnia need soft light a few hours before going to sleep.

The circadian rhythm for cortisol involves approximately ten bursts every day, with most of the pulsations occurring between 4 a.m. and noon. This rhythm served an evolutionary purpose, aiding our ancestors to be alert first thing in the morning, because safety required vigilance around dawn, when most predators were on the prowl. Ideally, during a six-hour period prior to 4 a.m. the absence of cortisol bursts allows for a restful sleep. In contrast, with chronic stress, cortisol levels do not level out, resulting in greater potential for insomnia.

A healthy circadian rhythm involves regulating body temperature. Insomnia can result from difficulty in regulating nighttime body temperature. Body temperature may actually increase at night when it should be going down, which may occur if the person fails to get any exercise in the daytime. During the early sleep cycle our body temperature should be in the process of dropping. Just before rising from bed our body temperature rises, as do cortisol levels.

A key way to keep body temperature cool at night, to achieve the deepest sleep, is to make sure the bedroom is cool. Warm bedrooms promote light and/or disrupted sleep. Taking a hot bath helps not only as a wind-down activity but also by raising body temperature while in the tub, which then drops sharply by bedtime.

Despite the folklore about alcohol promoting sleep, it suppresses slow wave sleep. Three hours after drinking alcohol, the induced glutamate blockade wears off, contributing to "sleep maintenance" insom-

> nia, also referred to as mid-sleep cycle awakening. This type of insomnia occurs during the middle of the sleep cycle, and it takes one to two hours to get back to sleep. Also, many people who suffer from insomnia resort to over-the-counter sleep drugs or physician-prescribed benzodiazepines. However, these drugs contribute to shallow sleep, minimizing slow-wave (deep) sleep, and lead to dependence and tolerance.

This chapter began by noting that self-maintenance comprises the first responders to mental health crises, as the critical stabilizers. Self-care behaviors must be addressed immediately, without which the traditional psychotherapy interventions may be fruitless.

Mental health studies have shown that lifestyle factors, diet, cognitive challenge, exposure to new environments, social engagement, and exercise on a regular basis are all associated with boosts of cognition and mood (Butler, Forette, & Greengross, 2004). I have used the mnemonic SEEDS— social, exercise, education, diet, and sleep factors—to help people remember to "plant" and cultivate them on a daily basis (Arden, 2014).

SEEDS:
S – Social
E – Exercise
E – Educate
D – Diet
S – Sleep

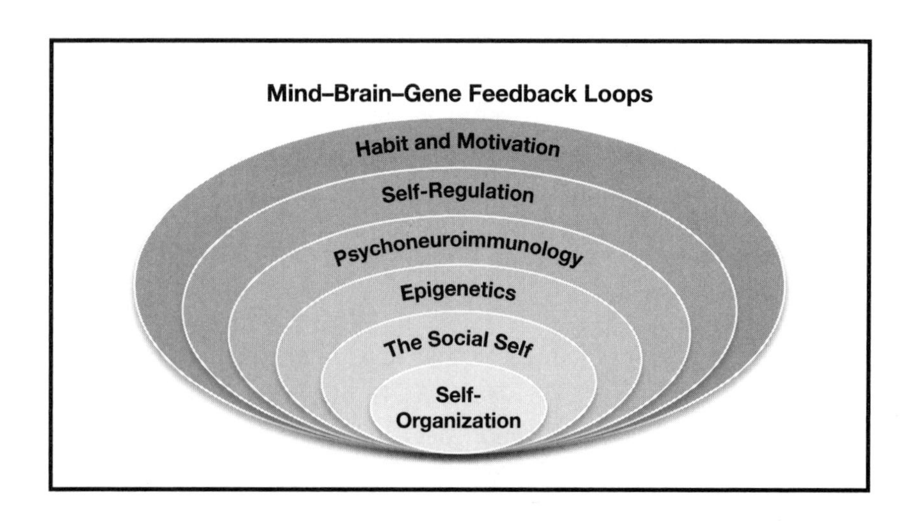
Mind–Brain–Gene Feedback Loops
Habit and Motivation
Self-Regulation
Psychoneuroimmunology
Epigenetics
The Social Self
Self-
Organization

SIX

Motivation, Habits, and Addiction

> We are what we repeatedly do. Excellence, then, is not an act, but a habit.
>
> —Aristotle

Sylvia always entered therapy enthusiastically. However, hope quickly dissipated after a few sessions, with her arguing that it was again not the "right fit." She terminated after the third session after detecting an inkling of pressure to change her habits of buying beyond her budget and excessive eating while glued to the TV. She claimed these were her "stress relievers." Though not quite a hoarder, her buying sprees got her out of the house and provided a rush of excitement at the moment of purchase. After slipping back into the dark hole of her "comfort cocoon," friends and family urged her to make another go of therapy. She would finally agree again, noting that her problems centered around stress at work.

Her therapists would discover that her job stress resulted from her supervisor's concerns about her lack of initiative and failure to turn in complete, neat, and timely assignments. When she sought the support of the employment assistance counselor at work, she would complain that "my coworkers demand too much of me," the same complaint that she had about her family. As she denied depression, anxiety, or other negative emotions, my gut feeling was that she was putting me on notice that I could not provide what she was looking for. In reality, she had no idea what she wanted or needed, and finding agreed-upon goals and a motivation to change was going to be a challenge.

SEARCHING FOR MOTIVATION

Clients are resistant to change partly because of the initial displeasure it brings. Many people, like Sylvia, come to therapy initially expressing motivation to improve their lives, and some are able to identify behaviors, mood, addictions, or thought patterns troublesome enough that they want to do something about them. However, too many people lose motivation and fall back into bad habits that they come to rationalize as viable coping tools—at least for awhile.

All animals possess behaviors critical for their survival, motivated by a reward-seeking system (Panksepp & Bivens, 2012). Our ancestors evolved reward circuits wired to take advantage of opportunities that were few and far between. When game and vegetable matter were available, they took as much as they could. Because food and other rewards were not constantly available, the reward circuit that added a spark plug to anticipate gratification so that we would seek out opportunities. This reward circuit served well during our evolutionary history to motivate our ancestors to seek out resources by hunting and gathering for survival, when complacency would mean death.

For those of our ancestors living in colder climates with harsh winters, the D2 dopamine receptors worked fine, not slowing down the drive to reward-seek even when rewards were plentiful. Those of our ancestors who lived closer to the equator did not have to jump at any opportunity because resources were rich and available all year long. Their D2 dopamine receptors were designed to slow down reward seeking when rewards were plentiful. With constant plenty, continuous reward seeking would have been maladaptive. In the developed world, those with the A1 allele of the D2 dopamine receptor are at a disadvantage, with people like Sylvia developing addictions to such habits as shopping, comfort food, and excessive computer use, social media, or television, and others developing addictions to drugs, alcohol, and gambling.

Adults with the A1 allele of the D2 dopamine receptor who have endured adverse childhood experiences tend to be impulsive novelty seekers who engage in high-risk behaviors, while those with the same genetics raised in a positive environment do not show these traits (Keltikangas-Jarvinen et al., 2009). People with the A1 allele of the D2 dopamine receptor may be more vulnerable to alcohol abuse and require a more nurturing environment to maintain mental and physical health. Adults who endured adverse childhood experiences also tend to have higher levels of dopamine and cortisol (Pruessner et al., 2004).

The Motivation-Pleasure Circuit

Sam was a hard-working insurance broker who said "most people die of boredom doing what I do." He said there is always a different combination of policies to discover, "like three level chess." While he was proud of his work habits, there was another habit he couldn't break: his obsession with spectator sports. It wasn't just that he was a fan of one or two teams, or had interest in one or two different sports. His addiction was to all sports that could be broadcasted on cable and radio channels, including pay-per-view events. While this may seem like an innocuous problem, his wife threatened to leave him if he did not quit. He admitted that even when they were out to dinner he would sneak into the bathroom to check his iPhone for up-to-the-minute scores of the games.

His drive to engage and maintain these bad habits is supported by a dynamic neurocircuitry, including the nucleus accumbens, a key part of the reward circuit and his salience network. Its proximity to the amygdala reflects how implicit memory system is intertwined with motivation, determining how much work to put into seeking a potential reward. The information stored in implicit memory regulates the release of dopamine from the ventral tegmental area, which signals the desirability, value, incentive, and salience of the reward.

Dopamine serves as a key player driving both Sam's ambitious work ethic and his addiction to spectator sports. Misconstrued as the pleasure neurotransmitter, dopamine is actually associated with the anticipation and motivation to seek the reward, rather than the reward itself. Thus, wanting is associated with dopamine circuits and working memory (Berridge, 2009). It provides incentive value to a stimulus and helps generate curiosity, interest, and motivation. In contrast, liking or enjoying something you've already attained involves opioid and cannabinoid circuitry. But the liking circuits also involve wanting. Though a variety of areas of the brain represent wanting alone, no regions represent liking without wanting. Many behaviors, including addictions, can transition into wanting without liking. Sam wanted to know all the scores of the games, though he found little pleasure in the knowledge. He stated that he could not wait to get home from dinner to watch all the games he taped. Yet he derived little pleasure from doing so. This distinction between wanting and liking represents an aspect of addiction. A person such as Sam feels compelled (wants) to watch, but he had grown to dislike it, and certainly himself afterward. This seemingly innocuous habit had taken over his life and pushed his wife away.

Not far from where I live in Santa Fe some of the Native American Pueblos had developed casinos to entice foolish people to wager their paychecks on the chance that they might quadruple their money. Ted was one of them. He said that each time he drove past the casino he felt a jolt of excitement, "not unlike a drug high." He complained of stress at work and poor health. When more stressed and unhealthy, he was more susceptible to the temptation of going to the casino.

Ted's dopamine neurons fire faster when the anticipated reward is greater. Driving by the casino cues these dopamine neurons to fire at a fast rate, but not as fast as when he walked into the casino. And the greater the rewards Ted expected, the greater the firing rate of his dopamine neurons. If the situation was particularly tempting, especially based on prior experience of rewards in similar situations, his dopamine neurons would fire particularly fast in anticipation. Like a Geiger counter approaching a radiation source, the closer he gets to the casino, the faster his dopamine neurons fire. As he drives past, just like the Geiger counter, the firing grows slower (Trafton & Gifford, 2008). The more dopamine neurons fire when an opportunity for reward is present (in this case the casino), the more likely the neurons in the nucleus accumbens initiate reward-seeking habits (in this case, going to the casino to gamble). The outcome of the reward seeking will determine whether the drive to seek reward in that situation will increase or decrease. When expected rewards are received, dopamine neurons fire as before the next time an opportunity arises. If the rewards are better than expected, firing will increase next time. If the rewards are not received, firing declines during future opportunities. In this way, dopamine neurons encourage reward seeking when opportunities are present and learn and adjust predictions about rewards as experience is gained.

Because dopamine triggers reward seeking (i.e., wanting, the drive to work for an anticipated quick benefit), dopamine is a strong driver of addiction. To become addicted you need to want not necessarily what you are addicted to—exaggerated wanting is enough. The activity of dopamine directed toward the nucleus accumbens generates activity in the reward-learning or habit-driving centers of Ted's brain. But unfortunately for Ted, casinos have learned to use the quirks of the brain's reward-learning systems to keep him coming back, such as slot machines making gonging noises when someone wins. These habit circuits amplify dopamine firing when rewards are unexpected, infrequent, and unpredictable. As a result, they encourage focused learning in situations when the brain cannot yet reliably predict when a reward will be received. By keeping the rewards large but unpredictable, casinos make the opportunities to gamble trigger unnaturally rapid firing rates. The intense craving Ted has for gambling involves an exaggerated firing rate in his dopamine neurons and a "rush" of excitement. To keep comparisons between natural rewards and these commercially amplified dopamine neuron responses to scale, Ted's brain was forced to start reducing dopamine neuron firing in response to natural rewards, making other life activities less motivating. Though Ted consciously may not enjoy gambling—like the lyric in the song, "the thrill is gone"—wanting that thrill lingers. He will continue to be drawn to

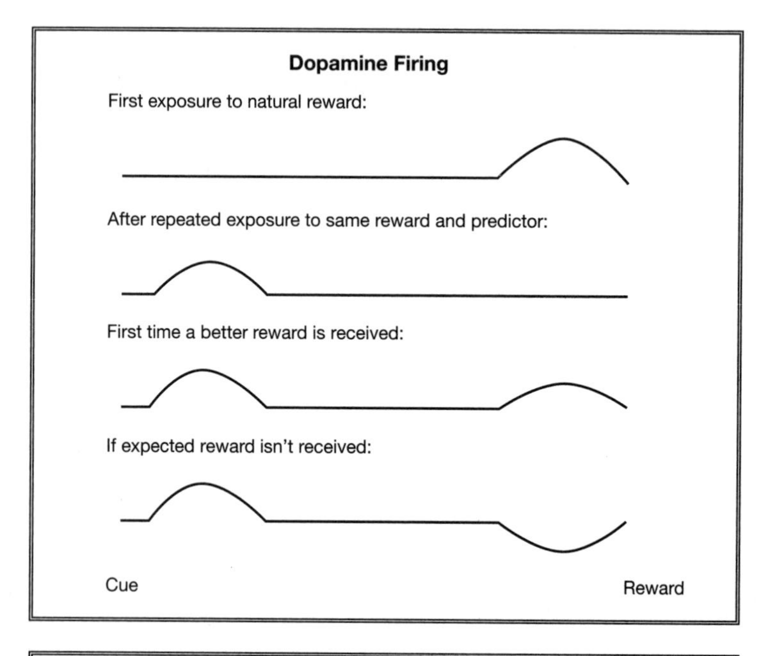

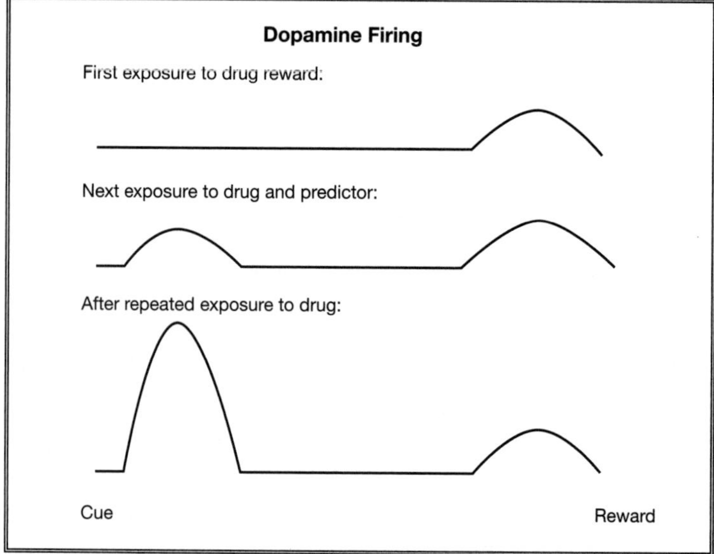

Figures 6.1a and 6.1b: Two possible sequences of dopamine activation leading to addiction. (based on Trafton, Gordon, & Misra, 2016).

the casino, as his learned gambling habits are initiated by the automatic response of his dopamine neurons to those cues he has associated with gambling wins in the past.

Because Ted's dopamine neuron firing rate reflect the potential for gambling wins, the faster they fire, the greater the expected win. Eventually, every association with gambling serves as a craving cue, generating a burst of firing fueling the anticipation of a jackpot and increasing likelihood that he will drive to the casino.

The challenge of inhibiting immediate gratification is that the activation of the reward circuit in the striatum can happen quickly, sometimes so quickly that it precedes completion of parallel decision-making processes in the prefrontal cortex. The prefrontal cortex is the region of the brain that employs conscious reasoning and can work to recall the past negative effects of gambling. Because reward-learning systems can make and act on decisions before conscious choices can be made, it is easier for Ted to decide not to drive to the casino than to resist gambling once he sees the blackjack tables, hears the clang of the slot machines, and has a few of the free drinks. These learned implicit cues predict potential rewards, so that his brain is sensitized to respond to them without conscious awareness. Once he developed the habit circuits in his striatum, frequent use made cells fire together and wire together associations between cues and gambling behaviors. The powerful drive for immediate gratification was hard to resist. His prefrontal cortex had a difficult time delaying gratification directed by his reward circuits, sending stop signals only after gambling behaviors had been started. Too often gambling won over his more carefully made long-term plans.

In the same way, Sylvia's reward circuit drove her to that second piece of cake. Although her executive network had to learn to foresee the long-term consequences, such as gaining more weight and becoming lazy and depressed, it was too slow to prevent her from eating. Though her executive and salience networks respond to dopamine, her executive network was focused on reasoning how to achieve better long-term rewards such as more energy, better mood, and improved health. Prefrontal decisions might align with the short-term reward-learning circuit decisions, for example, when considering the short- and long-term benefits of eating a nutritious gourmet dinner. But they might be at odds over decisions with short-term benefits but long-term negative consequences, such as eating cake. In these cases, Sylvia's choice will depend on whether the reward-learning circuits slow down long enough to wait for input from the prefrontal cortex.

> I can speculate that perceiving my smile of encouragement triggered a release of dopamine for Sylvia (Depue & Morrone-Strupinsky, 2005). Even when I offer a smile for a fraction of a second, followed by neutral comment, that quick reward increases her positive expectation so that she was more motivated to do things to improve her self-care. At work, too, the positive response to her new efforts are met with encouraging expressions because of her contributions.

Executive and Salience Network Integration

Priming Sylvia's motivation at work necessitated feedback between her executive network and her salient network's reward system. Her executive network kept track of her healthy intentions and goals, while interoceptive information was processed by her salience network about feeling full and satisfied, all of which influenced the dopamine-firing responses that provided valuation of the culinary options available to her. Her hippocampus helped recall episodic memories about the novelty of the decor and the creative menu and recipes. Her reward circuit uses this information to switch on dopamine release. The feedback to her prefrontal cortex and hippocampus ensured that she maintained access to information about all the rewards she most desired. The bottom line is that the same regions of the brain that provide inputs to the dopamine neurons also receive the outputs. In other words, she learned that the rewards are often more valuable if she inhibits the temptation for immediate gratification and acts toward achieving longer-term goals.

Beginning in the late 1970s a well-known study referred to as the marshmallow test illustrated how early ability to defer gratification predicts a lot about later success of the person. Children were presented one marshmallow and then told that if they wait until a little while later they could get two marshmallows, but only if they did not eat the one in hand immediately. The follow-up, years later, revealed that those who were more able to defer gratification (i.e., wait patiently with a marshmallow in hand) grew up to be more successful adults. Prefrontal cortex development facilitates the capacity to defer immediate gratification in the service of long-term planning and goal-directed behaviors. Prefrontal deficits, on the other hand, are associated with impulse control problems and affect regulatory problems such as anger, substance abuse, and relationship problems.

Of course, a person's success throughout life necessitates more than deferring gratification. The aspects of the salience network, including personal relevance, in terms of emotions, gratification, and meaning, are intertwined. The nucleus accumbens and the amygdala, which are in close proximity, play a significant role in his motivational system. The nucleus accumbens integrates incoming sensory information, evaluates information coming from the amygdala, and influences decisions made by the prefrontal cortex regarding whether to either "go" or "stop" on a possible motivated behavior (Hoebel et al., 1999).

> Because the success of psychotherapy depends on the clients' motivation to reach long-term goals over short-term gratifications, the incentives must be clear, as well as the personal relevance of their new behaviors that you encourage. If clients feel that their behavior results in positive outcomes, they will tend to repeat the behavior, which further strengthens the synapses driving dopamine firing and their motivation to engage in that behavior again.

Medium Spiny Neurons and Greater Gain

The capacity to respond to potentially pleasurable experiences that go beyond those immediately obtainable requires more than top-down control—bottom-up changes must take place, too. In other words, being able to defer short-term pleasure for long-term gains necessitates not only a well-functioning prefrontal cortex but also a refined subcortical reward system. To facilitate these bottom-up changes, the nucleus accumbens contains medium spiny neurons, so named because they have spiny shapes. Medium spiny neurons are part of two distinct circuits that respond to dopamine: the direct pathway, which includes neurons with D1 dopamine receptors, and the indirect pathway, which includes neurons with D2 dopamine receptors. The direct pathway acts quickly to motivate Ted to act now to gain immediate gratification: to stop at the casino on the way home instead of showing up for dinner with his family. The indirect pathway helps him resist the temptation to behave impulsively for short-term gain and immediate gratification. By toning down immediate gratification-seeking behavior, his indirect pathway enables him to consider his options and make better choices for long-term gain. In short, the indirect pathway puts the brakes on the quick-acting direct pathway.

The indirect pathway's ability to slow down immediate gratification-seeking decisions is experience dependent and takes learning. So how can Ted strengthen his indirect pathway? The indirect pathway will get stronger as Ted learns more options for achieving quick wins in his everyday life. Ted expanded the range of pleasurable activities in his life and learned effective, healthful ways to find comfort or relief in the face of problems, stress, and negative affect. The combination of increasing self-care, coping skills, and healthy pleasurable activities is what rehabilitation programs and Alcoholics Anonymous and Narcotics Anonymous support groups encourage. This is for good reason. Increasing exposure to small, natural rewards in day-to-day life leads to frequent activation of the D2 dopamine receptors in the indirect pathway, reshaping their responses to increasingly slow the reward-seeking decisions of the direct pathway. As a result, his indirect pathway brought his direct pathway under control, giving the prefrontal cortex and other brain regions time to weigh in on reward-seeking decisions. With this shift in decision-making control from short-term-focused direct-pathway circuits to long-term-focused prefrontal circuits, he sought gains when they were practical and in the interest of his long-term health.

Yet, increasing the range of pleasurable options and developing coping and problem-solving skills in the service of building his indirect (D2) receptors took time. Ted discovered that progressive shaping of goal-directed behaviors was easier when he broke up goals into small and obtainable successes. Ted's exposure to a greater variety of pleasurable options made him less vulnerable to immediate gratification. The more rewarding and varied his opportunities, the greater the strength of the indirect pathways, allowing his dopamine system to better weigh the overall value of all options—both immediate and long-term—and put the brakes on his impulsivity (Trafton & Gifford, 2008).

When Sylvia knew few ways to obtain immediate gratification, such as when she could recognize no benefits to attending a birthday party other than cake, her indirect pathway D2 receptor containing neurons became weak, and old habits were difficult to inhibit. However, when she learned social skills, such as that the people at the party become valued opportunities for approval and support, and problem-solving skills, such as that games and other party events became valued opportunities to overcome challenges, a party shifted from a bare environment, with nothing but the quick fix of cake to offer, to a rich environment filled with opportunities for near-term benefit. As illustrated in Chapter 3, genetic endowment does not determine behavior. What we have learned from the fields of epigenetics and neuroscience is that lifestyle and learning matters immensely. Behavioral health represents a continuum.

No Blank Slate

We do not start with the same vulnerability or potentiality. Neither do similar attachment experiences result in exactly the same styles of relating. Essentially, no one starts with a blank slate, contrary to what John Locke argued several centuries ago. People born with the A1 allele of the D2 dopamine receptor (with low D2 receptor availability) are more likely to develop addictions and have more trouble quitting

> once they become addicted (Comings & Blum, 2000). People with the A1 allele of the D2 dopamine receptor gene are thus more likely to become obese, develop gambling problems, and have trouble controlling urges for immediate gratification. They are also at greater risk of developing posttraumatic stress disorder after a traumatic experience. Because they tend to express fewer D2 dopamine receptors, those with the A1 allele of the D2 dopamine receptor have difficulty slowing the tendency toward immediate gratification seeking, including the drive to escape even mildly stressful situations.

When Sylvia's or Ted's D2 dopamine receptors (their indirect pathway) received increased levels dopamine, they became stronger and could restrain automatic-habit activities such as overspending for Sylvia and gambling for Ted. Strong indirect pathways supported control over their impulsive bad habits, so that newly formed good habits could dominate. However, when encountering overly tempting cues, such as shopping malls or casinos, the direct pathways are still strongly activated and may overwhelm the indirect pathways. This can prevent a client from inhibiting old habits.

With conscious and sustained attention, the development of new pleasurable habits blocked old habits stored in striatal systems. When more options were available, their D2 receptors became activated by dopamine, priming them to fire more easily and better inhibit reward seeking driven by the direct pathway (Dong et al., 2006). The total value of the rewarding opportunities that they experienced got translated into the strength and excitability of the indirect neurons (Trafton & Gifford, 2008). The richer their environment became, the more the indirect pathway slowed reward-seeking decisions, leading to greater consideration of long-term consequences. As they learned to recognize more rewarding opportunities, their indirect neurons were increasingly able to resist their old habitual immediate gratification-seeking behaviors. Indirect neurons are inhibitory and disfavor quick decisions to seek short-term gains regardless of long-term consequences, by putting the brakes on striatal circuits and their automatic habits. Essentially, by strengthening their indirect neurons though engaging in more varied beneficial activities, they reduced the tendency to act impulsivity and repeat bad habits.

> When the direct pathway is dominant, the salience network becomes more hedonistic and activity in the executive network is ignored. When executive network decisions are hijacked, they are in the service of explaining reward-seeking behaviors that have already been done, rather than strategic planning to guide good decisions. The brain's executive systems are relegated to making excuses for what the person has done. In doing so, the executive network rationalizes why, for example, gambling or overeating are not so bad. And the default-mode network repeats the story line of these rationalizations, perhaps even fantasies about the big win at the blackjack table. Through relapse prevention strategies that included contingency management, Ted was able to plan what he would do for pleasure instead of gambling. By developing stronger indirect pathways through D2 dopamine receptor activation, his executive network was able to switch from reactive crisis management to proactive strategic planning.
>
> Many clients state that they have a limited number of activities they enjoy. They report feeling stuck in bad habits and claim that their habits are hardwired. You can help them understand that the developed habit and the appreciation of pleasure is soft-wired by practice. From this understanding the client can build motivation to gain a wider range of gratifying experiences.

Without knowing the neuroscience that underlies habits, twelve-step programs have promoted expanding the range of pleasurable and healthy activities available for participants that do not involve alcohol or drugs. When participants cultivate a greater range of pleasurable activities, they become enriched not only by a variety life experiences but also by building the capacity to inhibit the drive toward immediate gratification. When previously old habits of drinking or drugging offered the only pleasurable experiences, the newly broadened range of go-to positive and healthy habits provides a new landscape of opportunity. The brain responds by recognizing this new wealth of possibilities, slowing decisions in order to consider all of its newly recognized options. As engaging in productive new habits becomes increasingly enjoyable, so does the motivation to continue a healthy course. However, breaking bad habits and establishing new habit requires moderating expectations.

Moderating Expectations

> Everything in moderation, nothing in excess
>
> —Aristotle

Optimizing motivation necessitates managing expectations. Because our brain develops expectations and predictions about rewards by varying dopamine activity, kindling motivation is not only a simple matter of finding a greater range of pleasurable behaviors. Dopamine activity changes constantly while we learn and update expectations based on the near-term outcomes of our behavior. Our brain recalibrates dopamine activity based on whether our decisions yield the rewards we expect. If the rewards were exactly as predicted, our dopamine neurons have done their job. If our choices do not result in the expected reward, our dopamine neurons will slow down to indicate we have made a mistake. Thus, our dopamine neurons both predict potential reward and provide feedback about the accuracy of these predictions.

The intensity and rate of dopamine firing both shape our behavior and guide decision making, especially when we encounter unanticipated rewards. For example, Pete had maintained sobriety for about a year when he met an old friend and his wife for lunch. As he walked into a café, he became elated to see an attractive woman named Sophia sitting with them, spiking his dopamine activity. The lunch was far more enjoyable than anticipated. A few days later he e-mailed his friends, saying, “I hope to see you all at the café tomorrow.” Of course he meant that they include Sophia. After his friend responded by saying, “Great, see you there,” his dopamine neuron activity shot up again. However, that evening when he entered the café and only saw his friend and his wife, his dopamine neurons dramatically slowed. His salience network dominated the other networks, not with positive feelings but with gut-level disappointment. He felt immediately like ordering a drink. As he turned to the wine list on the menu, his friend gave him a knowing and empathetic head shake. That social support was enough for Pete to feel temporarily soothed. Meanwhile, his executive network gained more balance with his salience network.

A few days later, his friend e-mailed to invite him to lunch at the café and noted that Sophia was invited too. Upon reading the e-mail his dopamine neurons fired up again. And when he walked into the café his dopamine neurons spiked again. As they all engaged in a lively discussion about traveling to Greece, he drifted briefly into his default-mode network, fantasizing about traveling with Sophia. Everything changed abruptly when she mentioned that she was excited about convincing her husband to go to Greece and take their kids there on a family trip. His dopamine neuron activity plummeted.

Our brain becomes temporarily shaped by a process of prediction and correction based on the accuracy of the predictions. Through trial and error, learning errors signal the difference between the prediction and the actual reward as we get feedback about what is really possible. Initially for Pete, dopamine neurons fired in anticipation of the potential reward of an intimate relationship with Sophia. As he learned to enjoy the relationship for what it actually offered, the dopamine firing rate became more successful at predicting the realistic value of friendship. After moderating his expectations, his motivation slowly increased to encourage the new friendship and trigger thoughts, emotions, and behaviors to cultivate it. This moderated effort allowed his reward-learning system to encourage reactions that supported enjoying the friendship with Sophia. By going beyond the search for highs and instead exploring the many gradations of positive experiences, everyday experiences can be enjoyed. Durable and sustainable motivation is particularly important when working to heal from addictions.

ADDICTION AND MOTIVATION

Pete had been struggling with alcoholism for several years. By the time he met Sophia he had convinced himself that it was time to get sober. His "bottom" amounted to losing his wife and being demoted at work with a stern warning. Those losses initially pushed him into recovery but were not enough alone to keep him there. Prior to recovery he had not developed regular pleasurable activities in his life. During stressful times, the instant gratification of alcohol was hard to resist. Alcohol had come to represent reward and relief from stress. The cues and reward-seeking responses occur so quickly that they are largely nonconscious. Dopamine-containing neurons in the ventral tegmental area will fire when tempted by the possibility of quick relief from alcohol even though it may be self-destructive in the long term.

Alcohol had previously tricked Pete's brain into valuing it more than other, healthier opportunities for reward. Alcohol had hijacked his dopamine reward systems and his nucleus accumbens to drive drinking behaviors while encouraging his prefrontal cortex to come up with rationalizations about its worth as a method of pleasure and coping.

Because our brain is essentially an organ of adaptation that addresses survival needs, living in a resource-deprived environment promotes brain activity that responds quickly to opportunities for rewards. This makes for an overly impulse-driven and reward-seeking brain that jumps at the chance for immediate rewards, especially when there are few to be had. When we enjoy a resource-rich life, we have no need to act on immediate rewards and can focus on long-term goals.

The classic example of this scenario occurred in the 1980s in the inner cities in the United States when cocaine in the form of "rock" became cheap and easy to score. Neighborhoods crumbled around crack houses. Currently in the United States the opioid epidemic was primed first by the overprescription of synthetic opioid pain medications and then by cheap and easy-to-obtain heroin. The resulting addictions have devastated many people and families.

The method of taking a drug affects the likelihood of developing an addiction: methods that deliver the drug to the brain quickly tend to be particularly overvalued by the reward system. For example, injecting an addictive drug delivers it directly into the blood supply and to the brain almost immediately. Conversely, drugs that go through the gastrointestinal track and interact with gut bacteria before being absorbed will lead to a slower, graded response of the brain to the drug. Of course, there are many variations within these extremes, such as drinking on an empty stomach or snorting drugs like methamphetamines or cocaine.

Also, how often the drug is taken affects the likelihood of developing addiction. Repeated use of addictive drugs leads to a loss of D2 dopamine receptors in the indirect pathway of the nucleus accumbens. A similar reduction of these receptors is seen with other nondrug addictions. Chronic consumption of high fat, salt, sugar, and junk food has been shown to lead to the loss of these receptors (Adams et al., 2015).

Suffering from the effects of chronic illness also leads to a variety of brain changes. People suffering from chronic pain suffer from the loss of D2 dopamine receptors in the indirect pathway (Martikainen et al., 2015). The use of pain medications may also reduce the number of these receptors. This downregulation of D2 dopamine receptors in the indirect pathway neurons occurs for a variety of reasons. Fortunately, the sum total of Pete's rewarding new experiences helped modify activity through the remaining D2 dopamine receptor neurons and lessened vulnerability to opportunities for immediate gratification. He built sustainable motivation by cultivating a wide range of healthy go-to behaviors that offered gratifying feelings.

When dopamine activity is moderate, the brain has time to recycle D2 dopamine receptors after use, and downregulation tends not to occur. In fact, there may be an upregulation if gene expression is shifted to make more receptors. However, when dopamine release is extreme, such as that caused by use of addictive drugs, the brain does not have time to recover. The extreme activation of all D2 dopamine receptors all at once leaves the neurons with no remaining unused receptors to respond. In contrast, with moderate activity neurons have time to recycle dopamine receptors. Normally, when dopamine binds to D2 dopamine receptors, the receptors change shape and cannot send another signal until they go through a recycling process. The receptor is taken inside the neuron and chemically treated so that it can return to a functional state. This recycling process is messy, with the loss of some receptors in the process. If loss of receptors outpaces the rate at which the neuron makes new ones, D2 dopamine receptor levels will decline. Moderate-size rewards stimulate moderate dopamine release, and a relatively small portion of the receptors go through this recycling process, leaving a large population of D2 dopamine receptors available to put on the indirect pathway brakes. In contrast, drug use surges dopamine release to the extreme; with overwhelming dopamine release the D2 dopamine receptor population becomes depleted. The person becomes less able to put the brakes on habits. In recovery those receptors come back over a period of weeks and months (Rominger et al., 2012).

Building on Strengths

The archaic tendency to segregate addictions from other psychological disorders fails to appreciate the habit circuits in the brain. People who suffer from anxiety and depression are more vulnerable to the development of addictions. In the 1980s there was a campaign to combat addiction with the simple suggestion to "just say no." Top-down willpower alone is not enough for most people to change bad habits into good ones. Certainly the prefrontal cortex networks are critical, but psychotherapy theorists have long identified the paradox of how when people apply their complete attention trying to stop a behavior, thought, or emotion they may actually increase them. Similarly, when they try hard not to have a panic attack, their mind paradoxically is on the lookout for any hint of a panic attack coming on. Any sensation, however subtle, such as a fluctuation in heartbeat or quickness in breath, tends to be amplified. Similarly, trying hard not to crave often leads to craving.

There are a variety of reasons why willpower alone tends to be inadequate. Through the prefrontal cortex circuits play major roles in initiating and inhibiting behaviors, there are limits to what a person can focus on at any one time. As the executive network, using working memory, attempts to focus on not using drugs, the thoughts of drugs nevertheless intermingle with decisions that may not be related to the substance the person is trying to avoid. This is because working memory is an active process; you cannot intentionally not do something without thinking intently about the thing you are trying not to do. With all that thought about drug use, it only takes a minor distraction to forget why you were thinking so much about drugs.

Multitasking, which produces working memory load, increases the activity in the anterior cingulate cortex to monitor and assess errors in learned behaviors (Weinberg & Hajcak, 2011)). But addiction can cause trouble activating the anterior cingulate cortex. This is why many recovery programs incorporate contingency planning and relapse prevention so that problem solving ahead of time can minimize working memory load.

> Developing healthy habits, especially during recovery, can be modeled, monitored, and reinforced by contingency management programs (Petry, 2000). Crucial to the success of monitoring and feedback include the following:
>
> - Making sure that the new target behavior is doable
> - Ensuring that the rewards occur for the new behavior upon completion
> - When possible, incrementally adding bigger rewards than smaller, as they provide more powerful impact
> - Making rewards for good behavior as opposed to punishment for failure

Changing the Habit Circuits

Changing bad habits into productive habits not only is challenging but also can be confusing. It is difficult to develop motivation to change because people typically feel worse before they feel better. For example, though Sylvia knew that she suffered from overeating, she gained temporary relief from stress by briefly activating reward circuits. Through her therapists' encouragement she had practiced eating nutritious food, but she felt more anxious before she felt less anxious. But when I explained that she would feel worse before she felt better, she shifted her expectations and gained motivation to persevere (Arden, 2015).

Comprising various aspects of the executive, salience, and default mode networks, the prefrontal cortex forms and modifies habits through three loops with subcortical areas, as shown in Figure 6.3 (Trafton, Gordon, & Misra, 2016). In addition to the planning, focus, and emotional involvement necessary to replace an old habit with a new habit, repeating and refining the new habit are critical for the development of healthy habits.

The lower loop includes circuits of the salience and default mode networks, including the orbitofrontal cortex, with projections to the caudate nucleus and from the thalamus. This loop is critically involved in habits related to reward-seeking and social restraint, affecting appetite and craving for food, drugs, and sex. Underdevelopment or damage to this area can

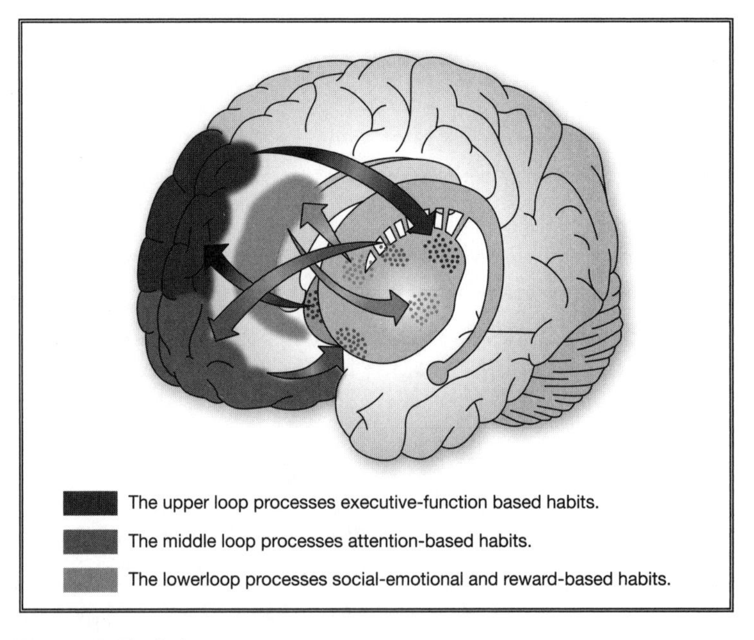

Figure 6.2: The habit circuits.

result in impaired empathy, distorted moral judgment, and risky decision making (Fuster, 2008). In fact, research has shown reductions in gray matter in the orbitofrontal cortex with drug-dependent people (Goldstein, Peretz, Johnsen, & Adolphs, 2007). When this loop is impaired a person tends not to gain from accurate gut level information.

The middle loop comprises the anterior cingular cortex, dorsostriatum, globus pallidus, thalamus, and then back to the anterior cingular cortex. Underdevelopment or damage to this loop can result in reduced motivation and goal-directed activity, loss of curiosity, and lack of interest in new experiences. This spectrum of deficits, along with the loss of concern for others, makes kindling this circuit important for people with depressive syndromes (addressed in Chapter 9).

Finally, the upper loop circuits of the executive network, consisting of the projections from dorsolateral prefrontal cortex to the head of the caudate nucleus and then back to the dorsolateral prefrontal cortex through the thalamus. This executive-network circuit involves planning and decision making. Underdevelopment or impairment in this circuit causes loss of a sense of memory for the future, failure to foresee contingencies, and inability to plan, prioritize, and engage in behavior that is flexible and preventative. Given that this loop is integrally involved in working memory, disruption here could result in losing the ability to organize the necessary activities to break bad habits.

All together, these three loops represent interrelated systems that can be strengthened to transform bad habits into healthy habits. The lower loop is involved in exclusion and the ability to suppress extraneous thoughts and manage affect, the middle loop maintains intentions to focus, and the upper loop maintains the object of intention.

KINDLING MOTIVATION THROUGH BOTTOM-UP AND TOP-DOWN APPROACHES

Most social-cognitive theories assume that intention to change is the best predictor of actual change. But as we all know, people do not necessarily behave in accordance with their intentions. For example, how people with an addiction respond to cognitive distortions may or may not influence their feelings and behavior. Changing how they think does not necessarily determine whether they will enter treatment or be successful in recovery.

There are thus limitations in a top-down approach. Cognitive reappraisal is a top-down strategy where the prefrontal cortex attempts to modify habit-based activity, including the amygdala (Goldin et al., 2012). Reappraisal engages the dorsolateral prefrontal cortex and the executive network to dispute negative thoughts, and ideally people can develop adaptive ways of responding to result in decreased amygdala and insular activity (Goldin et al., 2012).

Because direct top-down approaches can be complicated by bottom-up feedback loops, working with addiction requires accessing reward circuits. A multilevel approach may kindle activity in the salience network, where there are bottom-up and top-down circuits. Balanced by activity in the executive network to stay in the present, emotions, thoughts, and motivation can remain in sync. In this way, the prefrontal cortex moderates the enticement of the reward circuits with motivation for sobriety.

Helping people find the motivation to change has always been a challenge in psychotherapy. Some recent approaches have attempted to stimulate linking wanting with liking. For example, motivational interviewing attempts to expose the incongruity between what a person wanted and what came to be. It highlights the disagreement between actual behaviors and stated goals (Miller & Rollnick, 2012). Consistent with the acceptance commitment therapy concept of revealing the conflict between values and goals, these approaches attempt to help motivate behavioral change through accessing emotionally relevant circuitry as well as the reward system.

Change talk ("you should stop using opioids") inhibits activation in brain regions that respond to the salience of the addiction cues. Counterchange talk ("I can't live without my drug") activates multiple areas, including the insula and striatum, involved in the circuitry of alcohol dependence. Motivational interviewing may work by engaging and changing the neural network that supports the maladaptive habits. Perhaps for this reason, a meta-analysis of controlled clinical trials showed motivational interviewing to be a promising approach for treating disorders of alcohol, drugs, diet, and exercise (Burke et al., 2003).

Motivational interviewing proposes four critical elements:

- Expressing empathy
- Identifying the discrepancy between a client's actions and goals as a teachable moment
- Working with resistance
- Supporting the client's own reasons for change

Activating intrinsic motivation involves focusing on enjoyable and satisfying sources of motivation without immediate gratification. Building on intrinsic motivation, the natural tendency to seek out novelty, challenges, the capacities to explore, and to learn can form the foundation for sustainable change. By contrast, extrinsic motivation is characterized by secondary reinforcers, such as money, prestige, and praise.

SEVEN

Stress and Autostress

It is not stress that kills us, it is our reaction to it.

—Hans Selye

Tania had been healthy and quite satisfied most of her life. By age forty-two she had managed to earn an MBA and work her way up the corporate ladder, all the while feeling very invested in her role as a mother of two children and in a reasonably stable marriage. The challenges that she encountered along the way always seemed tolerably buffered by the support from a close group of friends. Then, in what seemed like a tsunami, much of her secure life seemed to wash away when her husband announced that he was having an affair and wanted a divorce. The very next week the CEO of her company propositioned her, offering a quid pro quo for further advancement. Feeling simultaneously assaulted by both men, she reached out to friends for support. But one of her friends was receiving chemotherapy for an aggressive cancer, and understandably the group's attention was on her. Nevertheless, she had the self-confidence to fight back without the immediate support of her friends. Soon after she filed a sexual harassment suit, the company unleashed their attorneys, smearing her repetition by claiming her divorce was evidence that she was pursuing the CEO. Meanwhile, her husband rationalized away his affair as a reaction to "her attempted affair" with the CEO. On top of all this, her children were beginning to act out during the emotional chaos.

Tania's health began to unravel soon after her husband moved out. Instead of maintaining her motivation to go to the gym and eat a balanced diet, she sat in her bedroom, staring at the television, eating pretzels. After suffering from insomnia, periodic hives, and an outbreak of herpes, she sought help from her primary care physician. What "bad timing," she told him, for her health to deteriorate and all these stressors to occur at the same time.

Was it bad timing for psychological stressors to coincide with Tania's bad health? Her stress systems, immune system, and poor self-care undermined her sense of stability. This chapter explores how Tania experienced the many permutations of the stress response. Immune cells have receptors for stress hormones and neurotransmitters, which allow stress responses to modify immune responses to prepare the body and brain for threats. This is a good thing over the short term. However, chronic and acute stress can morph into a self-organizing feedforward pattern of feeling stress when there is no actual threat. Like autoimmune disorders, when the immune system turns on and cannot turn off, autostress disorders represent the stress response system turned constantly on.

MIND-BODY CHAOS

Stress can lead to emotional chaos, which self-organizes into autostress disorders. To understand how, it is useful to explore the dynamic feedback loops between our homeostatic and stress systems. Homeostasis represents the capacity to reach physiological equilibrium at stabilizing set points to maintain a healthy body and brain. However, set points can be dysregulated by extreme or chronic stress and by elevated neurotransmitters and pro-inflammatory cytokines, which have the capacity to reset thermoregulation and glucose homeostasis. Because glucose is the main fuel of the immune system and the brain, poor diet, excessive stress, and sleep loss can disrupt glucose levels and dysregulate them.

When Tania encountered the onslaught of her husband's departure, sexual harassment by the CEO, and the threat by the company lawyers, the sympathetic branch of her autonomic nervous system released epinephrine and norepinephrine from her adrenal medulla into circulation to increase her respiration, heart rate, and dilation of skeletal muscles. Norepinephrine was also released from her locus ceruleus (in her brain stem), facilitating arousal and attention. All this transformed glycogen to glucose to facilitate energy to prepare her for action, commonly referred to as the fight-or-flight response.

For much of her life, Tania's autonomic nervous system responded effectively to temporary periods of stress. The "rest and digest" branch of her autonomic nervous system, referred to as the parasympathetic (meaning "above the sympathetic") nervous system, had put the breaks on her sympathetic branch after it was no longer needed. The parasympathetic nervous system encompasses the vagus nerve system (*vagus* means "wanderer"), which enervates many of the organs in the abdominal cavity (Porges, 2011). It is fundamentally interconnected to such body functions as heart rate and respiration and can slow down their activity as needed.

For at least the last few thousand years, people have cultivated activity in the parasympathetic branch without knowing of its existence. Through such methods as yoga, meditation, relaxation exercises, and hypnosis people have worked at achieving stress relief. Though many of the psychological aspects of these techniques are explored in Chapter 10, it is important to note here that the parasympathetic nervous system has a relationship with the immune system. It has been described as part of an "inflammatory reflex" because of its role in adaptive physiological reflexes that promote changes in illness behaviors (Tracey, 2002, 2009). It promotes the recouping energy expended during periods of stress and/or illness and after a stressful experience.

The vagus has many receptors for acetylcholine and oxytocin, which as noted in Chapter 2 play a role in the development of the social brain networks. Tania had previously enjoyed a strong social support system, and it had been a major part of her coping skills for much of her life. But with the social isolation combined with social stress, she had less access to her parasympathetic system without the social support. She had also previously maintained a balanced diet and a healthy sleep pattern and had gone to the gym regularly, but her motivation for those self-regulatory behaviors evaporated. As noted in Chapter 5, diet, sleep, and exercise also play a major role in mood regulation.

In addition to the overactivity of her sympathetic branch, know as the fight-or-flight system, her neuroendocrine system became dysregulated. Though not exactly sequential, the neuroendrocrine system normally kicks in to maintain alertness and energy for an extended period if the stressful situation persists. The neuroendrocrine system is also referred to as the hypothalamic-pituitary-adrenal (HPA) axis, because it includes the hypothalamus, which releases corticotropin-releasing hormone (CRH) and arginine vasopressin, which in turn stimulate the pituitary to release adrenocorticotrophic hormone (ACTH) into the bloodstream. Once ACTH reaches receptors on the adrenal cortex, the synthesis and release of cortisol facilitates more energy and alertness.

> Illustrative of the interdependence of the body's self-regulatory stress-response systems, "taking flight" recalibrates homeostatic systems. Exercise is well known for its role in modulating pain, producing the so-called runner's high, and decreasing stress. Beta-endorphin serves many functions, including as a buffer to stressful experiences by activating the parasympathetic nervous system and control over lymphoid organs. In so doing, it promotes circulation of innate immune cells (macrophages and natural killer cells), as well as anti-inflammatory cytokines. Serving as part of a negative feedback system, beta-endorphin helps bring the stress response back to a state of homeostasis, regulating CRH and ACTH by complex feedback loops to inhibit the activity of the HPA axis.

Stress: Immune System Disruptions

Tania's unchecked stress spurred by less parasympathetic activity led to her sympathetic nerve fibers firing excitedly. This meant higher than normal levels of norepinephrine. Extending into her immune organs, the release of norepinephrine in close proximity and directly into receptors in her lymph nodes, thymus, spleen, and bone marrow made her immune cells receptors "hear" the stress signals and reduce normal cell activity. Because most of her immune organs host norepinephrine receptors, stress and its messengers, CRH and norepinephrine, play a role inducing inflammation.

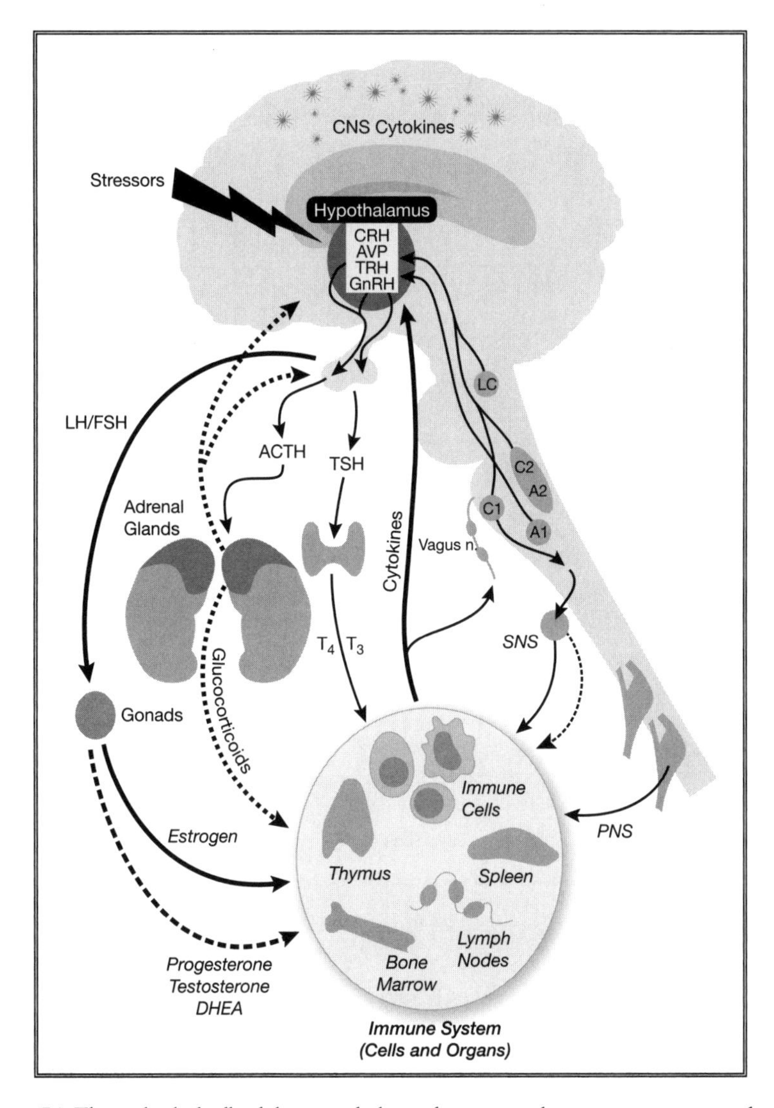

Figure 7.1: The multiple feedback loops underlying the stress and immune responses to threat.

One of the most spectacular illustrations of the dynamic changes in the immune response to acute stressors was measured in first-time tandem parachutists (Schedlowski et al., 1993). They had their blood drawn before the jump, during their time in the air, and after reaching the ground. The expected rise in epinephrine and norepinephrine occurred before and during the jump, followed by a spike in cortisol shortly after landing. Interestingly, natural killer cells were significantly elevated immediately after the jump but then decreased precipitately below starting levels one hour after the jump. This and other studies have shown that norepinephrine is the critical mediator of the immune effects of the fight-or-flight phase of the stress response. Acute and high levels of stress can impair the immune system through spikes of norepinephrine, which in turn activates natural killer cell activity only over the short term. In fact, a meta-analytic review of thirty studies showed that increased levels of the pro-inflammatory cytokine interleukin-6 follow brief stress (Steptoe, Hamer, & Chida, 2007).

The dynamic feedback loops between the stress and immune systems helped Tania's body and brain return to homeostasis most of her life. To ensure that her stress response returned to a balanced state, negative feedback mechanisms suppressed HPA activity when she did not need heightened alertness and energy. Key sites at the level of the hippocampus, hypothalamus, and pituitary exerted negative feedback to tone down her HPA axis after she met the challenge successfully.

Allostasis and Allostatic Load

> Do not judge me by my success, judge me by how many times I fell down and got back up again.
>
> —Nelson Mandela

Stress by itself is not harmful. In fact, having a totally stress-free childhood sets a person up for difficulty dealing with stress later in life. Mild doses of stress appear to result in subsequent stress resilience (Parker et al., 2012). For Tania, the critical factor was how much and for how long she experienced stress. Most important, what she did in response to the stress had a lot to do with why she developed health problems.

> Though extreme and/or extended stress makes it difficult to concentrate, mild to moderate stress can actually help us stay alert and focused. Moderate and fluctuating levels of cortisol energize us to engage in the challenge at hand. On the other hand, sustained and escalating stress ramps up levels of norepinephrine. The combination of high levels of cortisol, CRH, and norepinephrine further overactivates the amygdala, interfering with prefrontal areas, making attention and working memory falter.

Stress researchers pointed out that the term *stress* has major limitations. Popularized by Hans Selye and borrowed from engineering, suggested in the phrase "too much stress on this bridge," the term *stress* does not account for the dynamic adjustments that Tania made earlier in her life during challenging situations. A more dynamic model, referred to as *allostasis*, accounts for stability and change in response to unpredictable environmental demands (Sterling & Eyer, 1988). Because the stress systems are more dynamic than once assumed, many researchers have pointed out that the concept of allostasis accounts for the multiple feedback loops that provide the ability to maintain stability through changing situations (McEwen & Sterling, 2003).

The primary mediators of allostasis include the following:

- Stress hormones
- Immune system
- Neurological systems
- Mental operating networks
- Social demands
- Enhancing or inhibiting gene transcription
- Regulating brain-derived neurotrophic factor
- Up- or downregulating amygdala activity

- Targeting prefrontal systems involved in stress and emotions (Sullivan & Gratton, 2002)

The secondary effects of the primary mediators include the following:

- Serum high-density lipoproteins
- Cholesterol
- Glycosylated hemoglobin levels
- Waist-to-hip ratio
- Blood pressure (McEwen, 2002)

Whereas acute and chronic stress make it difficult to concentrate, mild to moderate stress helped Tania stay alert and focused. Healthy levels of cortisol energized her to engage in the challenge at hand. As a process of adaptation, allostasis regulated levels of epinephrine (adrenaline) and cortisol to promote adaptation to a short-term stressor.

The term *resiliency* has been used widely to denote durability, the ability to bounce back after a challenge. Like allostasis, the ability to maintain stability during challenges is critical for mental health. Resiliency involves a dynamic equilibrium between the activation of the sympathetic and its parasympathetic branches. A person can be aroused one moment, stressed at another moment, and recoup later. In dynamic equilibrium resilience depends on being able to shift states as the situation demands. Therapy can be thought of as means through which a client becomes resilient to moderate and episodic stress.

Allostasis requires adjusting physiological set points in response to stress. Like shock absorbers on a car, but far more dynamic, the opposing branches of autonomic nervous system make adjustments to challenges at any given moment. While the sympathetic branch releases epinephrine and norepinephrine, the parasympathetic branch applies the brakes on stress through the vagal nerve system. The HPA axis also flexibly adapts to the challenges at hand. While too much cortisol can suppress the immune system, too little cortisol can impair the ability to constrain inflammation.

As an allostatic hormone, cortisol maintains fluctuations throughout the day. It begins to rise as we get out of bed in the morning and continues to rise into midmorning, depending on the demands of the day, to ensure that metabolic energy is available for the challenges at hand. Optimally, it begins to drop around 4 p.m. until it reaches its lowest level around 2 a.m., during our deepest sleep. But if this circadian cycle is disrupted, as it was for Tania, and cortisol stays high into the evening, many health problems can emerge in addition to insomnia. Elevations in cortisol promote food intake so that glucose is available as fuel, and so she fell into her childhood habit of gorging on pretzels, which temporarily boosted glucose.

Cortisol increases appetite. Stress and overeating are also associated with the reward circuits. The nucleus accumbens is positioned close to the amygdala and the hypothalamus (associated with appetite, among many other homeostatic functions), so people tend to overeat when stressed. When stressed Tania was motivated to do whatever she could in the moment to make herself feel better, and overeating sufficed.

When she endured stress overload, the demands exceeded the gains from rest and recuperation. Sustained and escalating stress kicked up high levels of norepinephrine. The combination of high levels of cortisol and norepinephrine overactivated her amygdala and interfered with her prefrontal areas, making attention and working memory falter (Chajut and Algom, 2003). With her alarm response on too long, due to chronic stress, and epinephrine and cortisol levels elevated over many days, a condition referred to as *allostatic load* developed.

Allostatic load causes significant wear and tear on the body:

- Increased epinephrine
- Norepinephrine
- Pro-inflammatory cytokines
- Abdominal obesity
- Loss of bone minerals
- Immunosuppression
- Elevated blood sugar levels

- Atrophy and damage to the brain, especially the hippocampus
- Decreased vagal tone
- Acute-phase proteins
- Excessive cortisol (over the short term)
- Acceleration of atherosclerosis

With less social support and poor self-care as available buffers, chronic and acute stress overactivated Tania metabolic, cardiovascular, and immune systems. Initially her immune system was suppressed, especially with the activation of her HPA axis, with elevations in cortisol through its anti-inflammatory effect. Cortisol began shuting down all the body functions not immediately needed, such as digestion, immune functions, and metabolism. Meanwhile, it raised levels of adrenaline, norepinephrine, and cholesterol and increased blood pressure and heart rate. When prolonged and in the extreme, increased cortisol could have damaged the inner surface of her heart, as blood vessels tend to constrict and variability rhythm of the heartbeat decreases, all increasing her risk of sudden cardiac death.

> Ideally, a person achieves a dynamic balance in cortisol levels and lymphocytes (white blood cells). Cortisol can participate in killing lymphocytes through apoptosis (programmed cell death). The total number of lymphocytes in blood is inversely related to the amount of cortisol in blood. In other words, when cortisol is at its peak (in the morning), lymphocytes are at their lowest level, and when cortisol is at its lowest (at night) lymphocytes levels are at their highest.

Acute Stress > Ill Health > Poor Mental Health

Ahmed fled to Jordan after his neighborhood in Damascus was bombed. When he arrived at the refugee camp in Jordan, he knew no one. All his family had been killed. By the time I met him in a rehabilitation center in Amman, he was suffering from a wide range of health problems. When the nurse tried to give him his medications for high blood pressure and diabetes, he jerked away. He agreed to take them only after considerable encouragement from a fellow refugee.

Chronic and acute stress had increased his tendency to accumulate fat cells at the midline (belly), contributing to multiple health risks. Though he was gaunt, his visceral belly fat put him at risk for many disease processes, including metabolic syndrome, diabetes, cardiovascular disease, and other autoimmune and inflammatory disorders (McEwen & Wingfield, 2003; Steward, 2006).

The greater the number of stressors, the greater the cumulative effects of allostatic load. In fact, Ahmed suffered high allostatic load before the civil war. He had been living in fear that he would be killed by the secret police just like his father twenty years before. His social network revolved around working at a restaurant prior to the war. But it closed down into the first few months of conflict. With few options available to him to support himself, he joined the army, creating considerable cognitive dissonance. Tormented by memories of what happened to his father and seeing the atrocities committed by army compelled him to desert the army. He arrived back in his neighborhood a week before the bombing. Then when the bombing happened he speculated that the bombs were meant for him because of his desertion.

Chronic and traumatic stress have been correlated with the hyperactivation of the HPA axis and then the hypoactivation of this system. These deceptively contradictory effects have been the focus of considerable inquiry. Normally, there are variations in levels of catecholamines, cortisol, and CRH. Hypersecretion of CRH and cortisol and prolonged activation of the HPA axis in general have been associated with melancholic depression, panic disorder, obsessive-compulsive disorder, anorexia nervosa, and alcoholism. In contrast, hypoactivation of the HPA axis results in chronically reduced secretion of CRH, which may result in such health problems as fibromyalgia, chronic fatigue syndrome, and depression.

Hypocortisolism can lead to depression and metabolic syndrome, which are associated with risk factors for heart disease and diabetes. In combination with abdominal obesity, low HDL cholesterol, and high blood pressure, hypocortisolism dramatically tips the immune and stress systems into a dysregulated downward spiral.

Because allostatic load occurred much earlier for Ahmed than for Tania, the feedback loops that would normally maintain homeostasis broke down after sustained high levels of cortisol. Instead of signaling back to the HPA axis to terminate the stress response, it continued. For Tania the negative feedback loop worked quite well over the short term when all the systems were intact, much like a thermostat acting as a feedback mechanism to turn off the heat. As cortisol levels reached her hippocampus and hypothalamus the feedback mechanism kept the level steady.

Ahmed's thermostat, in contrast, broke down early. The excessive and prolonged high levels of cortisol probably damaged his hippocampus. Chronic acute stress not only led to atrophy in the hippocampus but also, combined with social stress, neurogenesis was blocked in the hippocampus and the prefrontal cortex (PFC) (Czeh et al., 2007).

Excessive cortisol had an opposite effect on his amygdala. Instead of a negative feedback loop, a feed-forward loop stoked up the hyperactivity of his amygdala, making him hypervigilant. This combination of the inability to turn of the stress response and the ramping up of the stress reactivity put him in an unbearable and confusing situation. Additionally, because his hippocampus functioned as a key part of his explicit memory system, his ability to remember insights derived from therapy sessions was compromised. It took considerable reassurance of safety and repetition of trust-building exercises for him to gain from therapy.

Poor health is both a consequence of chronic stress and a cause of undermined stress tolerance. To illustrate how this occurs, consider how Ahmed might have developed diabetes. He did not fit the typical profile of the millions of people in the Western world, who consume high amounts of simple carbohydrates, engage in little to no exercise, and gain extra weight. Instead, high stress pumped up his cortisol levels, signaling his cells to demand more glucose for fuel to deal with the constant danger. Chronically high levels of cortisol are associated with dampening the effects of insulin and are a risk factor for type 2 diabetes. Because insulin's principal job is to manage glucose levels so that cells can receive their fuel supply at optimum levels, this works fine when the stress is short-lived. But in high-stress situations that are chronic, as occurred with Ahmed, the endochronological and stress systems break down. During those periods of high and chronic stress the feedback loops are interrupted so that cortisol continues to be produced until the system becomes exhausted, resulting in hypocortisolism, inflammation, and type 2 diabetes.

> High stress activates genes in cells to increase glucose uptake. And high levels of adrenaline and cortisol raise blood glucose by triggering the liver to break down protein to convert to glucose. Also, high levels of cortisol contribute to excessive glucose, making receptors resistant to the effects of insulin. Thus, high levels of cortisol increase the risk of insulin resistance and thus type 2 diabetes.

AUTOSTRESS DISORDERS

Both Ahmed and Tania developed anxiety disorders, but the patterns of how the anxiety were organized were quite different. In the mental health system we have identified these patterns of emotional chaos as diagnoses associated with common clusters of symptoms. For Ahmed trauma played a major role—Chapter 8 picks up his and similar stories. Though for Tania the prognosis was less severe, she did have trouble regaining the resiliency she once enjoyed. She landed a better job after the company awarded her a generous settlement for her suit, but she continued to suffer from anxiety. Her stress systems turned on and stayed on, and her overreaction to subsequent stress contributed to an anxiety disorder.

Because anxiety disorders feed on the stress response systems they may also be referred to as autostress disorders. Like autoimmune disorders that hijack the immune system that turn against the body instead of protecting it, autostress disorders turn on the stress response systems so that Tania felt there may be an imminent threat when there was really just a pattern of false alarms. In other words, anxiety develops when the stress systems are turned on too often and signal danger when there is none.

Autostress disorders can develop with unbalances in the top-down patterns of activity in the brain. In the exaggerated bottom-up pattern the amygdala hijacks the PFC so that Tania constructed all sorts of semi-plausible reasons why she felt fear. Those anxious feelings emerge even without an identifiable stressor to feed an autostress disorder. Anxious feelings and thoughts spurred on unchecked by excessive amygdala activity are associated with anxiety disorders, whereas top-down inhibition of this bottom-up overactivity is fundamental to reducing anxiety. Accordingly, the therapeutic reduction of anxiety is associated with increased PFC activation and decreased amygdala activation (Schienle, Schafer, Hermann, Rohrmann, & Vaitl, 2007). In other words, how she cognitively framed a situation determined whether it would be anxiety provoking or not.

While the amygdala is associated with threat detection, it does not register fear. Fear is a conscious emotion; the amygdala is involved in the nonconscious detection of threat and contributes to behavioral and physiological responses (LeDoux, 2016). These nonconscious ingredients to the emotional feelings of fear and anxiety are not reducible to them or the primary source. The motivational states are interpreted by the mental operating networks, which construct the conscious feelings of fear and anxiety.

Sensitization/Priming

Autostress disorders develop through cross-sensitization, which involve increases in intensity each time stress is encountered. Sensitization, also called priming, occurs when repeated stress increases exponentially in magnitude. In other words, to add to the phrase "cells that fire together wire together," each time they do so, they do it much more quickly, and it takes very little to set them off. For Tania this meant that hearing a neutral tone in her new boss's voice instead of the usual upbeat tone triggered anxious feelings.

> Sensation/priming involves the compensatory adaptation of the amygdala-HPA axis, which can occur long after stress so that ACTH and cortisol are set a lower 24-hour levels. Hormones such as arginine,

vasopressin, and the catecholamines act synergistically with CRH to increase the stress response. When the person experiences a new emotional stressor, the prior priming of the amygdala-HPA axis responds through an even higher level of ACTH, making this primed system hyperrespond to stress.

Because prior acute and/or chronic stress increases the release of pro-inflammatory cytokines, subsequent stress can potentiate or prolong their release (Johnson et al., 2002). Cross-sensitization occurs through multiple feedback loops, such that one condition potentiates another. Ahmed, who had previously experienced trauma, developed an autoimmune disorder. As he became acutely stressed again, his autoimmune conditions worsened, increasing his anxiety and depression.

Stress that transitions into autostress disorder is associated with a number of factors:

- Decreased vagal tone
- Elevated fasting glucose
- Increased pro-inflammatory cytokines
- Acute-phase proteins
- Increased cortisol

Exposure Versus Avoidance: Balancing Side to Side

Tania wanted help to avoid "triggers" for her anxiety. Though avoidance gave her momentary relief, it was actually making her anxiety worse over the long run. Avoidance of even slightly stressful experiences represents one of the principle factors that contributes to autostress disorders. Avoidance of what is feared primes the stress system to fear anything associated with it and avoid it at all costs.

Prior to the settlement with her company, she alternated among many forms of avoidance, such as escape, procrastination, and safety behaviors. With escape behaviors, she immediately removed herself from anxiety-provoking situations, such as meeting new people and staying clear of social gatherings. She procrastinated to the last minute about anxiety-provoking situations that she had no choice but to attend, arriving late to meetings. And she used safety behaviors to buffer herself from assumed danger by always sitting with coworkers who demanded very little engagement from her.

All these avoidant behaviors sensitized her right PFC and amygdala. Because avoidance is consistent with a right-hemisphere bias for anxiety disorders, she tended to feel overwhelmed generalized fears and worries. She had trouble identifying the practical details and short-term goals, which were processed by the left hemisphere. When socially phobic, she at times felt "speechless" because Broca's area, which is associated with expressive speech, is located in the left frontal lobe.

This side-to-side frame of reference involves the distinction between the big picture and details: while the right hemisphere tends to see the big picture, the left sees the details. With right-side overactivity, she was overwhelmed with the totality of her situation and became anxious. The right side is less able than the left to inhibit the overactivity of amygdala. With more activity in the right amygdala, there tends to be more threat detection and anxious memories. In contrast, more activity in the left amygdala is associated with less anxiety and positive memories. With an overactivation of the right PFC and amygdala and an underactivation of the left PFC, fears add to a feeling of timelessness, in one overwhelming gestalt.

Exposure to what is inappropriately feared is fundamental to therapy for all anxiety disorders. Associated with approach behaviors, exposure accordingly kindles the left PFC. Even though it did not feel right to her immediately, facing her fears eventually helped balance the activity in both hemispheres. Helping her understand this paradox motivated her to resist avoidance behavior.

However, exposure too quickly, referred to as flooding, is like jumping into the deep end of the pool from the high dive before learning to swim. One of the reasons that flooding can be countertherapeutic is because it does not fully utilize the incremental talents of the left PFC and tends to skew the dominance to the right PFC. As a result, the person feels overwhelmed with the totality of anxiety. Because of these limitations, exposure must be managed incrementally so as to expand the comfort zone instead of constricting it. While flooding tends to abruptly constrict the comfort zone, failure to incrementally experience challenges gradually shrinks it.

Exposure exercises must do the following:

- Generate incremental steps slightly out of the comfort zone
- Break down goals to doable chunks

One of ways that exposure works to neutralize anxiety is facilitated by the release of beta-endorphin through its anxiolytic effects. As an endogenously produced opioid compound with analgesic properties, it is activated in response to physical pain, temperature change, conditioned fear, and even social conflict. Beta-endorphin is released into the bloodstream from various endocrine sources in response to stress after about fifteen to twenty minutes. There are higher concentrations of beta-endorphin in the morning than the afternoon.

Beta-endorphin is coreleased along with adrenocortical hormone (ACTH) but is momentarily blocked by ACTH at the common receptor sites. The therapeutic effects from exposure in part result from the more rapidly decaying ACTH, which is displaced by the delayed beta-endorphin anxiolytic effects minutes after the exposure (Carr, 1996). Runners refer to a "hitting the wall" when they feel that they can run no more. The wall evaporates with a "second wind" made possible by the release of beta-endorphin, which provides the means to deal with pain and stress. This means that if we tolerate a period of increased anxiety for at least twenty minutes, we will be rewarded by less anxiety.

Exposure to stress-provoking stimuli until the fear and anxiety fade is thought to involve three structures: the amygdala, hippocampus, and cortex (Hartley & Phelps, 2010).

1. The amygdala is needed for threat detection, as well as for relearning a new safety cue. These implicit memories of threat are essentially emotional habits that change slowly through relearning in therapy.

2. The hippocampus is needed for learning the context of safety.
3. The newly formed explicit memories code into the cortex realistic associations to overshadow the previous false alarm memories.

The context must not be too narrow and restricted but, rather, flexible and generalized to as many real-life experiences as possible. Failure to broaden the context results in greater risk for relapses that occur outside of the new learning environment.

Exposure activates the ventromedial PFC and hippocampus to inhibit the overactivity of the amygdala, suppressing the detection of threat when there is no threat. In fact, the thickness of the ventromedial PFC is correlated with the success of exposure-based therapy.

Top-Down, Bottom-Up, and Side-to-Side Integration

While some therapeutic approaches emphasize bottom-up processes, cognitive behavioral therapy and, in the United Kingdom, cognitive bias modification offer a top-down approach. Neither the top-down nor bottom-up perspectives offer a complete explanation for the development of anxiety or how to neutralize it. An integrated approach neutralizes anxiety by dampening bottom-up neural circuits together with activating top-down circuits. Normally, the PFC and amygdala work together to ramp up or tone down the stress response systems, so that Tania could rise to the challenges when appropriate and calm herself when there is no realistic danger.

As noted in Chapter 1, evolution ensured that the tracks going up from the amygdala to the PFC are stronger than those going down from the PFC to the amygdala. Survival from danger has always had precedence over any other situation. For Tania the tracks going down from her PFC to her amygdala were temporarily weakened, and thus her ability to distinguish what was really dangerous from what just felt like danger contributed to anxiety. The goal of therapy was to strengthen the tracks going down and weaken those going up.

The top-down processes have a side-to-side dimension, which involve how she cognitively framed her situation. She tended toward excessive right hemisphere activation and not enough left. Cognitive behavioral therapy has a long tradition of identifying of cognitive distortions, which are essentially negativistic exaggerations. A typical list includes black-and-white thinking, overgeneralizations, catastrophizing, all-or-nothing thinking, jumping to conclusions, and so on. All of these frames of reference are consistent with a heavy emphasis on right-hemisphere activation. In other words, there are no shades of gray, no gradations, no subtle details—instead only extremes, all-inclusive and anxiety provoking. Damping down the stress response system necessitated orchestrating not only activating the top-down but also left PFC neural circuits that were underactive. This meant cultivating an appreciation of the shades of gray, the details, and approach behaviors.

Generalized Anxiety

When the stress system stays abnormally active without stressful situations, the emotional chaos can self-organize into different types of anxiety disorders. Some anxiety disorders become generalized, while others become focalized into anxiety centered on a specific stimulus or situation.

Mona suffered from the generalized type. For as long as she could remember she always seemed tense and worried. As a child, she shared concerns with her mother that her alcoholic father would do something or say something disturbing. Well after she married Matt and had children, she sought therapy to help rid herself of feeling tense. Matt was a mild-mannered insurance salesman who was consistently reassuring to her during what she called her "worry frenzies." He often gave her back massages to dissolve her constant feelings of tension. She complained that "there is something wrong with me! Maybe I have an anxiety gene?"

Generalized anxiety disorder (GAD) differs from focalized anxiety such as panic or phobias by a persisting feeling of tension and uneasiness, as if something terrible could happen at any moment. GAD involves the breakdown in the dynamic equilibrium within the autonomic nervous system, with a tilt away from the parasympathetic branch that provides rest and relaxation in favor of the sympathetic branch that generates arousal and the emergency response. For Mona the negative feedback loops that maintained flexible balance between the branches became dysregulated. A positive feedback loop was set in motion so that she felt overwhelmed with free-floating anxiety and excessive worry when there was no identifiable stressor.

One method to rebalance the two systems involves a variant of slow and deep breathing. While inhaling shifts the body toward the sympathetic branch, exhaling stimulates the parasympathetic branch. To illustrate the difference, I asked her to consider that when she laughs or sighs she exhales. In contrast, when she gasps there is an abrupt inhale that activates her sympathetic branch. By reestablishing the dynamic balance between the two branches she learned to establish flexible access to her parasympathetic branch.

Variations in vagal tone correspond to how well a person can calm down during or after a period of high stress. Higher vagal tone provides better self-soothing capacity, more reliable autonomic responses, and a greater range and control of emotional states than does low vagal tone. Mona developed GAD in conjunction with an underdeveloped vagal brake, which resulted in the hyperstimulation of her heart in response to stress. When her vagal brake was off, her heart rate was up and she felt like she was in a constant state of emergency. Because her vagal brake was weak, her alarm response persisted well after encountering stress, with elevated epinephrine and cortisol over many days.

Approaches that activate parasympathetic activation include somatic exercises, include a focus on breathing:

- Hatha yoga
- Mindfulness
- Contemplative prayer
- Relaxation training
- Guided imagery
- Self-hypnosis that emphasizes breathing

- Autogenic training
- Body scanning

Relaxation methods that incorporate a focus on breathing should not be engaged in as a form of avoidance but, rather, as methods to cultivate easy access to the parasympathetic nervous system. However, relaxation-centered therapy renders, at best, an incomplete positive outcome. To address the full range of GAD symptoms therapy must address worrying.

Worry as Cognitive Avoidance

Avoidance associated with GAD is not as obvious as it is with focalized anxiety. Avoidant behavior with GAD kindles the worry circuits. Mona worried to avoid uncertainties. There is a positive feedback loop between excessive worry and feeling anxious, so her worries contributed to overreacting to stress, and stoked up more anxiety and yet more worry. With free-floating anxiety she worried because she felt uneasy, but worrying perpetuated the tendency to worry even more, as a feed-forward loop within her default mode network.

Mona's worry loop was fed by her intolerance to ambiguity and uncertainty. Combined with poor problem solving and positive beliefs about worry, her cognitive avoidance was an attempt to think herself out of discomfort. Ironically, she only found more uncertainty and discomfort, because worry is both consciously and unconsciously self-reinforcing.

She often asked her husband for reassurance that each of her worries were unjustified, but this only fed her insatiable worry loop. Meaning well and conforming to the natural tendency to comfort, he provided a short-lived fix to resolve her specific worries. Reassured that a specific worry was not worthy of concern, she moved onto yet another worry.

Often hypervigilant, Mona scanned her environment for potential danger, avoiding rather than approaching stressful situations. With intense anticipatory anxiety, she tied up the cognitive resources of her executive network and used her default-mode network to reflect on worries. Because of her limited capacity to tolerate neutral novel stimuli, developing a well-functioning orienting response was critical. She needed to learn to react appropriately to novel and potentially ambiguous situations.

Because she tended to select threatening interpretations of ambiguous stimuli, therapy focused on increasing her tolerance for ambiguity. By learning to tolerate ambiguity through exposure to and development of an appreciation for ambiguous situations, she built a stronger executive network. During her reflective moments with her default-mode network, she learned to appreciate the subtleties of ambiguous situations. She learned to call these skills characteristic of maturity and "wisdom."

> Because the amygdala functions as an orienting subsystem for the rest of the brain, alerting other systems to gather information, it serves as part of an integrative system that crosses categorical boundaries such as emotion, motivation, vigilance, attention, and cognition (Whalen & Phelps, 2009). By itself the amygdala cannot tolerate ambiguity (as highlighted by the still face paradigm described in Chapter 4). It contributes to the overestimation of negative outcomes to neutral (ambiguous) stimuli, so that the person is tormented with free-floating anxiety.

One of the methods that helped her learn to tolerate ambiguity can be referred to as "boring the worry circuit." We used a variant of the cognitive behavioral therapy technique called "scheduling worrying time," then took it a few steps further. In preparation for the "worry hour" she carried around a notebook so that as worries occurred to her throughout the day she wrote them down. Then between 5 and 6 p.m. each evening, she opened the notebook and devoted the entire hour to reflecting on all the worries. Instead of reasoning about the worries, she was asked to reflect on how each of them represented an uncertain and ambiguous situation. Eventually her worries became boring or, alternatively, ambiguities that she found interesting. This method helped build her PFC and hippocampus circuits through exposure to these ambiguities. Eventually she learned to accept various degrees of ambiguity as harmless and interesting.

The goal was to observe the worries uncritically and label them as merely worrisome thoughts without trying to solve or "cognitively correct" them. Borrowed from the contemplative traditions, this method promotes detachment of the thought from the feelings of anxiety. In metacognitive models such as acceptance commitment therapy and dialectical behavioral therapy, transforming an anxious thought from the thought-emotional "fusion" to "thought diffusion" is illustrated in the change in statement, "Oh, that thought makes me anxious," when there is no distance between an anxious thought and the feeling of anxiety, to "Oh, that's an anxious thought." In the second statement the thought is merely a thought, nothing to get anxious about. Such worrisome thoughts may flit in and out of the default-mode network without priming the worry loop. However, when they elicit the emotional charge from the salience network, so that the worries "feel" personally relevant without the participation of the executive network, the worry loop feeds forward. The process labeling of affective states by the executive network reduces anxiety by weakening the dominance of the connection between the default-mode network and salience network.

Not only is being able to tolerate ambiguity important for the management of worry, but as Mona learned it also reflects maturity. Moreover, being curious about ambiguity cultivates complexity of thought and greater cognitive skills.

The Transition Between Cognitive and Metacognitive Models

Cognitive Behavioral Therapy	Metacognitive Models
Rationale: control	Rationale: relinquish control
Cognitive restructuring	Thought diffusion
Breathing retraining	Observe and accept
Interoceptive exposure to lessen fear and avoidance	Interoceptive exposure with acceptance of internal cues
Exposure to lessen fear fear and avoidance	Exposure to achieve life values and goals

Focalized Anxiety

Autostress disorders that become fixated on a particular situation or stimulus can transform the chaotic feelings into a panic disorder or a phobia. Attention focalized on physiological sensations, as is the case with panic disorder, or a specific object or situation such as a phobia, generates fear and avoidance to feed the disorder.

Katy grew up in a family with a father who had a hair-trigger temper and a mother who walked on eggshells in an effort to keep from "setting him off." She described him as a rageaholic. When he launched into a tirade, everyone immediately tensed up, trying not to be the target of his rages. By the time she was in her fourth year of marriage with Scott, Katy was confused and agonized that she tended to suffer from a panic attack when there was even a hint of mild disagreement. She knew quite well that Scott was almost the opposite of her father. Yet, she wondered why she panicked when discussing even the most subtle emotionally provocative subjects with him.

Through interoceptive fear she interpreted stimuli inside of her body such as a rapid heartbeat as a signal that she was in danger. Her focus on these sensations seemed to fuel her panic symptoms. In contrast, exteroceptive fear (phobia) relates to stimuli outside of the body, such as if she feared objects or situations like strangers, enclosed spaces, or in Katy's case, with Scott during slightly stressful conversations. Yet, Scott never gave her any reason to fear him. When growing up her interoceptive fear of body sensations and exteroceptive fear of her father potentiated each other. If had she responded to either exteroceptive or interoceptive fear with avoidance, she created the very symptoms that she tried to avoid. She had interoceptive fear of a rapid heart rate, hyperventilation, a dry mouth, sweating, and butterflies in her stomach, which signaled that she was in danger.

Her primary care physician prescribed a benzodiazepine that "helped only for a little while," and then her anxiety "got worse." Not warned about the phenomenon of anxiety sensitivity associated with benzodiazepines, she incorrectly assumed that she needed stronger and stronger doses. Due to the tolerance and withdrawal effects of benzodiazepines, she became prone to panic.

> Overbreathing leads to excessive dissipation of carbon dioxide and a condition referred to as hypocapnic alkalosis, making blood more alkaline and less acidic, which involves the following:
>
> - Peripheral vasoconstriction, leading to tingling in the extremities
> - Oxygen binding tightly to hemoglobin, resulting in less oxygen released to the tissues and the extremities
> - Vascular constriction, so that less blood reaches the tissues
> - Dizziness, light-headedness, and cerebral vasoconstriction, leading to feelings of unreality
> - Abnormally sensitive carbon dioxide receptors in the brain stem, which may lead to a "false alarm" of suffocation
>
> Normally, carbon dioxide helps maintain the critical pH level in the blood. With lower pH, the nerve cells to become more excitable, so that thoughts race and associate with feelings with anxiety and panic.

Katy was referred to my anxiety therapy group. There she learned to begin to identify how her fear of her own bodily sensations was the problem, not the sensations themselves. She constantly monitored her body for any physical sensation that might "warn" her that an attack was on the way, which made things worse. In addition to tense conversations with Scott, she learned that her previous attempts to avoid anything that might stir up those bodily sensations, such as jogging or running up the stairs, contributed to a panic attack. Sometimes her panic attacks seemingly came out of nowhere; she tried to control the conditions in which these sensations occur, which only to led to the presumption that she gained temporary control.

The interoceptive exposure techniques helped her habituate to her own bodily sensations so that they no longer triggered panic. She and Scott came in for a few couples counseling sessions, during which she was taught to apply the techniques. Interoceptive exposure involved teaching her to systematically desensitize herself to those bottom-up physical sensations. During the interoceptive exposure she was encouraged to embrace these physical sensations so that they could eventually become part of a richly textured and interesting life.

The Slow and Fast Track: Balancing the Executive and Salience Networks

Katy's threat detector circuit had become primed to overestimate danger more quickly than most people. Evolution as endowed us with a threat detection system that requires no thought. The amygdala functions as a key part of the threat detector and the salience network. It can receive sensory information before the executive network. In other words, we can detect a threat before we think that there is a threat. This "fast track" to threat detection works fine when there are real threats. However, Katy's fast track to her amygdala was on too often, and her slow track needed to speed up. Slowing down her fast track while speeding up the slow track helped her develop the infrastructure of affect regulation. In the fast track scenario, when Katy heard a hint of anger in Scott's voice she either froze to speechlessness or left the room in panic. She felt frightened before she had time to determine if there was reason to be frightened.

Fast and Slow Track to the Amygdala

The fast track to the amygdala is made possible through the circuits coming directly from the thalamus. The slow track includes circuits going from the thalamus to the cortex and hippocampus before reaching the amygdala.

To buttress her executive network, Katy developed the ability to observe the hypersensitivity of her salience network. Initially, she flinched and her heart quickened, but then she eventually learned to ask for clarification from Scott about what he was addressing. This maneuver bought her time to gather her thoughts, kindling her executive network.

One of the ways that helped Katy speed up her slow track was to conceptualize layers of thought: automatic thoughts, assumptions, and core beliefs (Arden, 2015). Automatic thoughts represent a superficial thinking layer of the slow track as the "first thought to pop up" when beginning a tense discussion with Scott. Those thoughts were primed by an implicit memory bias when her fast track detected threat-based feeling states. Her executive network had been hijacked and attempted to confirm the feeling of threat to construct fear. To speed up her slow track at the automatic thought level, she learned to interrupt the fast-track impulse with a pause, to allow her executive network catch up, to become aware of the present moment, and to realistically appraise the situation. She asked herself, "What is really happening, and what can I say that is constructive?"

Her assumptions had been mini-theories she held about her inadequacies that she reflected upon within her default-mode network. She was encouraged to follow up by the actual behavior that steps toward her fears instead of avoiding them. Despite the fears, acting "as if" she could handle the situation gave her the opportunity prove to herself that the fears were unfounded.

The third layer involved her core beliefs, which represented existential life descriptors that she framed as having been irreparably damaged by being a "child of a rageaholic." She reconstructed a description of herself as a durable survivor, encouraged by a shift from a global/passive (right PFC) to a detail/action (left PFC) approach to develop a transformed identity.

In sum, recovering from anxiety involves balancing mental networks that orchestrate the side-to-side, top-down, and fast and slow tracks. Like autoimmune disorders, autostress disorders represent the inappropriate turning on of the self-protective systems to detect threats when there are none. The resulting emotional chaos we call anxiety develops when a person shifts to avoidant behavior, fails to maintain self-care, and/or develops autostress disorders. Whether anxiety is generalized or focalized, recovery from anxiety neutralizes the overactivity of the threat detection and stress systems. While generalized anxiety disorder, panic, and phobia can develop over time, trauma can occur abruptly, as explored in Chapter 8.

EIGHT

The Trauma Spectrum

> Although the world is full of suffering, it is also full of the overcoming of it.
>
> —Helen Keller

At no other time in written history have there been more people on the move as refugees, with many traumatized by war, famine, and genocide. The numbers are estimated to triple in the next thirty years, with climate change causing whole island nations to disappear and coastlines to move inland, all contributing the social chaos and trauma. These historical changes, combined with the ongoing trauma experienced by people who endure such disruptive events as domestic violence, rape, and auto accidents, make the next few decades extremely challenging for health care professionals.

Not all people respond to life-threatening situations in the same way. Some people are quite resilient, while others develop a wide spectrum of chronic anxiety, including posttraumatic stress disorder (PTSD) and depressive disorders. All the feedback loops addressed in previous chapters, including epigenetic, immune system, attachment, and social factors, play major roles in undermining or supporting "self"-organization and resiliency for people who endure traumatic experiences.

Linda had at least a few risk factors for PTSD before entering an abusive marriage. She suffered years of abuse from Steve before I met her. And it wasn't her idea to seek therapy. Only after her third visit to the emergency department for bruises and a broken nose did she agree. The staff knew what was happening, despite her denials that she was the victim of domestic violence. I met her at 3 a.m. while on call to the emergency department. We arranged an immediate referral to a domestic violence program where I previously served on the board of directors. Perhaps it was because her broken nose was so obvious or my firsthand information about the program, but she showed minimal resistance to the placement.

Risk Factors for PTSD

- Greater distress before and after the trauma
- Poverty and low socioeconomic status
- Previous or current psychological disorder and poor affect regulation
- Family discord and/or insecure attachment
- Cognitive disengagement at the time of the trauma and dissociation involving depersonalization and derealization
- Early and repeated trauma

After two weeks at the shelter she came to see me for follow-up as an outpatient. She noted that as soon as she began to feel understood by her peers in the support group, she became less hesitant about disclosing the extent of the beatings. Up to that point she had sought help for a wide spectrum of health problems, including obesity, type 2 diabetes, and high blood pressure, but not the domestic violence. She had had assumed that Steve would "take mercy" on her because she was too ill to abuse.

As a daughter of an alcoholic who beat not only her mother but also her, she seemed to have a marriage initially that was "somewhat normal." In fact, Steve did not hit her until well after his "verbal tantrums" went on largely without protest. She explained that she spent much of their marriage "walking on eggshells," because "there was no telling what would set him off."

A person with a greater internal locus of control may be more durable and resilient in the face of trauma. In response to the first incident of verbal abuse and domestic violence she would have immediately set firm limits or gotten out of the marriage. In contrast, Linda had an external locus of control. She was more vulnerable to develop PTSD because she possessed fewer resources and less confidence. She doubted that she could deal with the traumatic stress and recover afterward.

With apparently less PFC development supporting affect regulation, her capacity to make and then follow through on practical goals was compromised, especially when she was stressed. In fact, she did not believe that her behaviors could prevent or protect herself from abuse. In short, she developed learned helplessness that contributed to not only depression but also the cycle of violence.

THE BREAKDOWN OF THE SYSTEMS

As noted in Chapter 7, our species evolved a multilevel stress response system. Given that we live in an extraordinarily complicated social environment, many of the stressors we encounter are social. If the danger we encounter is from other people, our first line of defense is the social engagement system. We talk, implore, and even beg for mercy if our life is in danger from others. If those maneuvers fail, we can revert to the fight-or-flight response. And if fleeing or fighting off a threat is not possible, and if the situation is hopeless, we may revert to the most primitive response of shutting down, becoming immobilized. People who have been traumatized may use any or a combination of these defensive responses.

Stages of Defense

1. The social engagement system: the myelinated vagus
2. Fight or flight: the sympathetic nervous system and hypothalamic-pituitary-adrenal axis
3. Immobilization: freeze, collapse, and feigned death, in two stages—
    - Freezing in terror
    - Paralyzed—shut down: total submission, trancelike dissociation

Traumatic events provoke defensive behaviors that may work over the short term but not the long term. Linda's neuroendrocrine system, sympathetic branch of the autonomic nervous system, and immune system became dysregulated in response to acute and chronic stress. When his rages went on for hours, she could not regain allostasis because the sustained effects of these hyperalert states led to the breakdown of her self-regulatory systems. Even when he was not posing imminent risk, she had difficulty falling or staying asleep and concentrating during the daytime. Her default-mode network activity centered on replaying traumatic autobiographical memories or fears of the future.

When Linda detected an immediate threat from Steve, her amygdala signaled the locus ceruleus to release norepinephrine and her adrenal medulla to release norepinephrine and epinephrine (adrenalin). Both neurotransmitters promote immediate hypervigilance, quicken heartbeat, and tighten muscles to prepare to flee from danger. Her quickened thoughts, centered on whether Steve would hurt her at that moment, were processed by her cortex, and memories about whether he did so in similar contexts before were processed by her hippocampus. If she determined that a real threat was imminent, her orbital frontal cortex disinhibited her amygdala to activate her neuroendrocrine system.

Whereas her sympathetic activity occurred immediately, her neuroendrocrine system (hypothalamic-pituitary-adrenal axis) added sustained alertness and energy, by her hypothalamus releasing corticotropin-releasing hormone within seconds of hearing his angry voice. Approximately fifteen seconds later her pituitary released adrenocorticotropic hormone, and within minutes her adrenal glands released cortisol to increase her metabolism to utilize glucose stores for more energy to deal with the danger at hand. Sometimes her stress systems were not amenable, and in response to the abuse she shut down into what she called her "deer-in-the-headlights" mode.

STABILIZING AND PREVENTING PTSD

There are significant differences in the potential consequences of experiencing a traumatic incident and experiencing several traumatic events. As described in Chapter 3, the greater the number of adverse childhood experiences endured by a person, the more health problems and the more vulnerability to subsequent trauma. Linda endured several adverse childhood experiences with complex trauma. In contrast, one discrete trauma, like a near-fatal auto accident, though not simple, is relatively less complicated.

Take Michael, for example, who suffered from the residual effects of a horrific auto accident that totaled his car and killed all drunk teenagers in the car that hit him. He became hyperconditioned to stimuli associated with automobiles. Because our brain abides by the evolutionary mandate of self-preservation, strong associations with a trauma are exceptionally resistant to extinction (Milad & Quirk, 2012). Accordingly, Michael felt compelled to avoid cars altogether.

Immediately after the accident Michael's norepinephrine, adrenalin, cortisol, and pro-inflammatory cytokine levels were all elevated. Nightmares and hypervigilance plagued him during the first few weeks following the accident. Had he seen a psychologist initially after the accident, the therapeutic effort would have been directed toward stabilization so that less traumatic memory was encoded by lowering of elevated stress chemistry. Elevated levels of these neurochemicals, associated with the development of PTSD, serve to maintain intrusive thoughts and nightmares, as well as ramping up anxiety by unleashing the amygdala.

This potentially debilitating combination of factors emerging post-trauma demonstrates that therapy should be directed toward stabilization and prevention of PTSD. With this in mind, critical incident debriefing employed just after the trauma is largely considered countertherapeutic. Heightening stress neurochemistry by a premature review of the traumatic details accelerates the encoding of traumatic memory and actually primes the development of PTSD.

Therapeutic interventions should include those that stabilize and calm, to minimize the risk of developing PTSD. One of the principle goals of psychological first aid is to depathologize and normalize the initial hypervigilance and nightmares. If people are not forewarned, when they do experience those symptoms they become alarmed and frightened and overreact, making the symptoms worse.

A variety of feedforward factors encode the stress response into an enduring chaotic pattern that misreads safe conditions as danger. In fact, if one month after a traumatic event the levels of three neurochemicals are elevated—norepinephrine, high cortisol in the evening, and proinflammatory cytokines—there is a higher likelihood of transitioning from acute stress disorder to PTSD.

Neuroimaging studies involving PTSD symptom provocation have identified consistent findings (Bremner, 2002):

Reduced Activity

- Left hemisphere
- Dorsolateral PFC
- Hippocampus
- Broca's area

Increased Activity

- Parahippocampus gyrus
- Posterior cingulate
- Amygdala

These factors inhibited the capacity of Michael's PFC for problem solving and rational behavior. Meanwhile, he was plagued by heightened startle response, vigilance, insomnia, flashbacks, intrusive memories of the accident, and increased fear conditioning, so that he even avoided riding in his car as a passenger.

The symptoms of trauma create a feedforward loop, driven by neuroplasticity, intensifying over time. Even when Michael watched movies that featured a scene with a car chase, the trauma felt like it was happening *now*. Each time this happened, it led to further priming and the potential for more flashbacks.

The three therapeutic phases applied to trauma are as follows:

- Phase 1—stabilization immediately after a traumatic event; interventions offer "psychological first aid" to stabilize, calm, normalize, and orient.
- Phase 2—integration of the dysregulated implicit and explicit memory systems to promote affect regulation and regain allostasis and "self"-organization.
- Phase 3—posttraumatic growth directed toward long-term recovery with resiliency, based on a sense of meaning and direction in life.

Balancing Bottom-Up and Right/Left

The positive relationship among amygdala activity, hyperarousal, and startle response and the reduction in activity in the medial PFC represents a failure to exert appropriate top-down inhibition. The amygdala functions as the driving force (the accelerator) to defensive reactions, whereas the medial PFC regulates (applies the brakes) to the amygdala. This check-and-balance system is often impaired in people who have been traumatized, so there tends to be more acceleration and less braking for stress. Because the amygdala is part of a threat detection system that the cortex assembles into the conscious feeling of fear, Michael tended to feel threat when there was none. To make matters worse, the sustained high levels of cortisol impaired his hippocampus thermostat function, making him prone to hypervigilance and false positives for danger.

One of the goals in therapy was to shore up Michael's PFC and its ability to dampen the hyperarousal and reexperiencing. This effort put more emphasis on the left PFC through establishing and working on concrete goals, over the tendency to withdraw, associated with the right PFC. The structure and follow-through on doable goals supported his self-efficacy and helped stabilize and prevent the development of PTSD.

Long-Term Dysregulations

Had Michael not sought out therapy, the unaddressed residual effects of the trauma would have contributed to the dysregulation of his immune system. That is what happened for Linda. Distant and persisting trauma may also result in inadequate cortisol; its natural anti-inflammatory effect evaporated, and she suffered from chronic inflammation and autoimmune diseases. While too much cortisol can suppress the immune system, too little can impair the ability to constrain inflammation. Chronic stress, depression, and PTSD are associated with hypoactivity of the cortisol system.

Complex trauma breaks down many of the homeostatic feedback systems. In addition to the cortisol receptors on the hippocampus, part of the hypothalamus called the paraventricular nucleus provides another negative feedback mechanism that inhibits further release of cortisol. As Linda endured acute stress and repeated traumas, these thermostat functions broke down. She was less able to shut down the stress response, so that even minor stress felt overwhelming.

High levels of stress for extended periods forces the body to do the following:

- Utilize elevated levels of glucose
- Increase pro-inflammatory cytokines
- Increase acute phase proteins (Blandino, Barnum, & Deak, 2006)

The stress-induced autonomic-inflammatory reflex contributes to the following disorders:

- Metabolic
- Vascular
- Autoimmune (Bierhaus, Humpert, & Nawroth, 2006)

Acute and chronic stress, combined with depression, increases the risk for impaired immune system functioning, coronary heart disease, myocardial infarction, chronic pain syndromes, type 2 diabetes, and dementia. Acute stress contributes to activation of pro-inflammatory cytokines (Harbuz, Chover-Gonzalez, & Jessop, 2003). Accordingly, Linda suffered from chronic inflammation combined with elevated catecholamines, making her feel ill, depressed, and anxious at the same time. In response, she failed to maintain positive self-care behaviors, which further undermined her health and impaired her central nervous system, undermining her ability to manage stress. The cumulative effects contributed to more depression and exacerbated her autoimmune disorders.

These feedforward loops involving stress, depression, and autoimmune disorders illustrate how traumatized people not only tend to become stressed and depressed but also are more vulnerable to feel ill because they are stressed and depressed. The mediators of allostasis operate as a nonlinear network, so dysregulation of any of these networks leads to allostatic load.

Memory Integration

After Linda left Steve, her capacity to make use of safety-related information was obscured by intrusive trauma-related thoughts and feelings. Because at times she suffered from a fragmented sense of self and the loss of a cohesive experience of time, her traumatic past could instantaneously surge into the present, hijacking her sense of safety. Her emotions were fragmented from her cognitions, which all felt too overwhelming to put both into context together.

The therapeutic reconsolidation of her memory systems was key to her regaining a sense of self and capacity to regulate her affect. Because her implicit memory system was out of synch with her explicit memory system, surges of anxious feelings were triggered by sights, sounds, and smells. Her explicit memory system, a function of her cortex and hippocampus, was overwhelmed by the onslaught of dysregulated anxious feelings coming from her implicit memory system.

Implicit memories are more resistant to change and the passage of time, in contrast to explicit memories, which are modified over time. Implicit memories are difficult to update by purely verbally based approaches because they occur spontaneously through exposure to sights, smells, sounds, and emotions. The more complex and extended the traumatic experiences, the greater the tendency to code in a range of unregulated implicit memories.

The interaction between the explicit and implicit systems is influenced by the intensity of emotion. Because explicit memories depend on the hippocampal-frontal memory system, during periods of intense emotion they can be overtaken by amygdala-driven implicit memories. Intense emotion associated with trauma, combined with reduced hippocampal processing of explicit memory, tends to favor greater formation of implicit memory. Implicit system can contort the explicit system to code in threat-based memory. Avoidance of the traumatic memory cues maintains symptoms of trauma, whereas emotional engagement with traumatic memories neutralizes them and is critical to recovery.

Therapy with Michael involved helping him incrementally focus on the implicit memory content of the flashbacks without suppressing them. These exposure-based exercises included sitting in a car, driving it, and eventually driving past the location of the accident. He managed to depotentiate the timeless qualities of the implicit images and sensations so that they became linked with a spatial and temporal context monitored by his executive and salience networks. Consolidation of the implicit system with the explicit system frames traumatic memories in the past, where they belong, and that he is safe in the present.

The process of exposure and reconsolidating the implicit memories within the explicit system must be repeated multiple times to facilitate neuroplastic change. Through rehearsing the newly consolidated explicit memories with new narratives, Michael learned to say, "I felt that way right after the accident when the paramedics pulled me out of my mashed up car. But I'm in no danger now." Through slow exposure to the previously frightening implicit memory cues, he brought them gradually under control so that they no longer triggered flashbacks. He simultaneously applied meaning, context, and realistic thoughts to seemingly chaotic sensations and emotions.

So-called hot spots represent brief moments of emotional intensity associated with flashbacks, such as when Michael drove into the neighborhood of the accident. These moments of dysregulation between the contents of implicit and explicit memory systems identified the focal points needing reconsolidation. His sustained attention to the implicit memory cues, such as the feelings of driving his car in that neighborhood again, strengthened his coping skills and enhanced his prefrontal circuit's inhibitory control over the amygdala, diminishing flashbacks. Integration of his memory systems was framed by new narratives, such as "I will drive defensively, and that is the best I can do." Therapy focused on reconsolidating traumatic memories from the accident into feelings of realistic self-efficacy.

This shift in activity within his salience network allowed his previously fragmented implicit memories to transform into an integrated sense of stability. The cognitive self-representations available for narrative expression either in the moment through executive network or in imagination through his default-mode network contained the disturbing emotional feelings that came up. Instead of reverting to overwhelming flashbacks, he could stay in the present and calm himself.

Memory integration necessitates the following steps:

1. Assessing the current capacity of the explicit memory system to determine the pace in which integration of the implicit and explicit systems is possible.
2. Identifying the cues that trigger flashbacks.
3. Priming the integration by orchestrating a "safe emergency" within the window of tolerance to apply an exposure exercise to the implicit memory cues.
4. Ensuring that integration is encoded through multiple processing channels, including somatic grounding.

Normally, the realm between explicit and implicit memory can be hazy. Especially for Linda, implicit memories seemed surreal and horrifying during flashbacks. Episodic (explicit) memory is relatively rational, due to its reliance on cortical-hippocampal networks. In contrast, emotional and procedural (implicit) memory is irrational and more dependent on subcortical networks. The dynamic interface between the contextually organized time-dependent episodic memory and the generalized, irrational, and timeless emotional memory was dysregulated. The integration of explicit and implicit memory slowly brought meaning and positive emotions to her life.

Though fragmented, her explicit memories of the trauma needed to be made cohesive and realistic so that she could explain the implicit feelings as they came up. For example, when Linda said, "My head is beginning to feel dizzy," she was able to reframe the experience by saying, "This is how I used to feel when Steve was about to hit me. But he is not here now, and I am safe." Therapy rebuilt her explicit memory system abilities of context and time and integrated them with implicit memories of the trauma, which had up to that point felt timeless. Because the trauma-related episodic memories were fused with negative valence, the feeling tones needed positive valence (Levine, 2012). For Linda it began with the belief that she was beginning to feel safe and in control of her own life.

Co-constructing a new narrative placed past events in the context of her new experiences within the supported environment of the domestic violence program. Though the process took considerable time, the benefit of living in a therapeutic community provided constant support and encouragement for her salience network to reconfigure a viable sense of self. She was slowly able to integrate past experiences into a cohesive sense of self which, though unfair, could be accepted as understandable. Therapy helped her construct a new adaptive narrative of being a survivor instead of a broken person. Ahmed, introduced in Chapter 7, also gained from a therapeutic community. He joined others who were suffering from the traumatic complexities of civil war. Like Linda, he developed a narrative consistent with conceptualizing himself as a survivor instead of damaged for life. But unlike Linda, he was plagued by secondary emotions.

SECONDARY EMOTIONS

Ahmed used his executive and default-mode networks to draw up memories from his explicit memory system to evaluate how the events could have been prevented and the implications to his self-worth. So-called secondary emotions associated with the episodic memory are not experienced at the time of the traumas. They include regret about risks he had taken or sadness and remorse at the loss. These emotions may involve shame and guilt about his failure. He felt tremendous shame for joining the army that supported a regime that killed his father. He finally deserted after witnessing more atrocities and the fear of being forced to participant in them in the future.

Secondary emotions can torment trauma survivors for their entire life. When I first heard about the tortured guilt and remorse experienced by my grandfather Missak, I was speechless, and my heart ached. He had witnessed the murder of his first wife and child while hiding in the woods. The Turks were gathering all the young Armenian men for labor camps, the army, or execution on the spot. Women and children up to that point had been spared. At that moment the Turks brutally shifted to outright genocide. The year was 1915. Though Missak managed to escape with his life, part of him never left that spot. The tormenting guilt and deep sadness he carried would periodically erupt in rage. Decades later he encountered two Turkish men in Pershing Square in Los Angeles. Trying to rise above his grief and reconcile the painful memories, he approached them, greeting them in Turkish. When they asked if he was Turkish, he could not resist responding, "No, I am one of the few Armenians that survived the genocide." One of them shook his head in disgust and said, "What genocide!? You Armenians need to get over the past." Missak had to be restrained from attacking the man. Later Missak explained: "It was as if it was all happening again."

His story permeated my family. But he was not the only one to have experienced trauma. I learned that, like him, many of my maternal and paternal relatives carried with them similar horrific stories. The next generation and mine to follow carried an undercurrent of survivor's guilt.

This complicated mix of sadness and horror is made worse not just by a few men in a park in Los Angeles but by the continued organized program of denial of the genocide by the Turkish government.

For my grandfather, just hearing Turkish spoken in an angry voice instantly brought him back to that horrifying day. His implicit memory system contained information that had been obtained from sensations from the traumatic events, including the sound of a Turkish accent spoken in anger, which triggered bodily sensations such as increased heart rate, temperature, and pain. These implicit memories cued flashbacks that occurred immediately and involuntarily. This is why he said, "The past is always there . . . lingering, ready to bite. And it comes with a vengeance!" The raw, poorly organized, implicit situational reminders of the trauma, such as the physical sensations and horrifying emotions, never left him. Superimposed on the erratic implicit memories was a unintegrated layer of guilt and deep sadness.

MANAGING THE WINDOW OF TOLERANCE AND OPPORTUNITY

Like many people who have experienced complex trauma, Linda tended to be in a state of biphasic hypoarousal and hyperarousal. The cues for her flashbacks were often fragmented implicit memories that appeared sporadically and at times "out of nowhere." On the other hand, at times she felt numb, blunted, and in a surreal world. A window of tolerance was available between these extremes. The task of therapy was to operate within that window. Implicit memory cues are difficult to neutralize without deliberate exposure in therapy. Yet, they are best approached through well-managed exposure within this window of tolerance to build the explicit memory system so that the implicit memory cues can be tolerated and integrated.

Prior to her time in the domestic violence program and therapy, Linda alternated between feeling overwhelmed with intense emotional states, such as sadness, despair, fear, and shame, overreacting to the world around, and feeling distant, numb, and as though little part of her had died. This oscillation of hypervigilance at one extreme and avoidance and emotional blunting at the other extreme undermined her sense of self-efficacy and control. Her window of tolerance represented the space between these two extremes, in which she could best achieve integration and positive therapeutic outcome.

Linda's state-based implicit memories were cued by specific breathing patterns, muscle tensions, gastrointestinal sensations, heart rate, and the tone of my voice—any or all of these sensations served as implicit cues for a flashback. As with people with panic disorders, interoceptive exposure exercises served as access to these somatic sensations. The increase of implicit memories and the decrease of context-dependent explicit memories opened the gate for flashbacks triggered by sights, sounds, and smells, experienced as timeless threat.

Because her implicit memory system was nonconscious, it could be

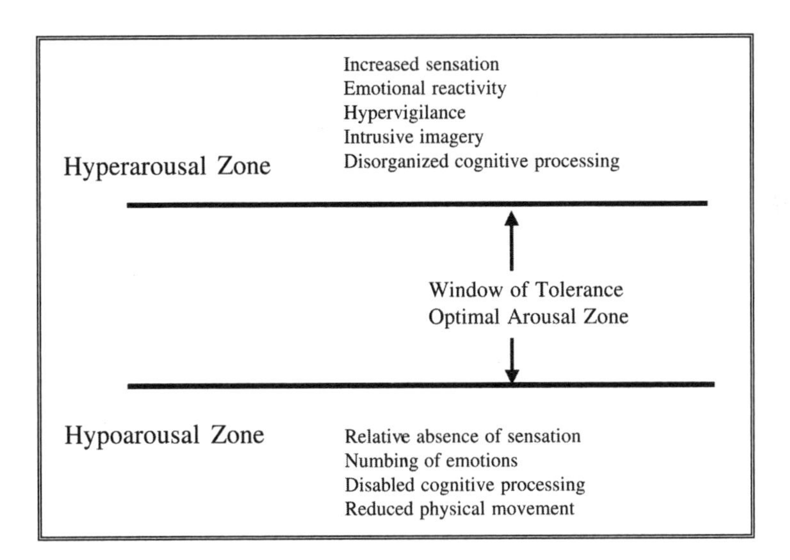

Figure 8.1: The Window of Tolerance

triggered by the fast track to her amygdala. The sensory information such as sounds went directly to her amygdala from her thalamus to signal a potential threat. Primed to activate the false-positive threats by elevated levels of cortisol and norepinephrine, her amygdala overwhelmed her hippocampus and PFC. Consequently, her executive network, which would normally restrict threat detection to realistic danger, was off-line. Therapy entailed balancing bottom-up and top-down circuits within the window of tolerance.

Integration of her implicitly generated arousal levels were carefully managed to help her affect stay within the therapeutic window so that she was not too hyper- or hypoaroused. Simultaneously, we made sure that the "safe emergency" of the exposure exercises utilized the moderatere level of arousal to maximize neuroplasticity. If her arousal levels were too low, traumatic memories would not be accessed. On the other hand, if arousal levels were too high, she would have been so overwhelmed with the traumatic memory that the effort would have become countertherapeutic and retraumatizing. She alternated between feeling too much or not enough emotion. The goal was to help her get back into the range of tolerance, back to dynamic equilibrium.

Because her hippocampal and PFC networks may have been compromised, therapy strengthened her executive network, carefully integrating implicit information such as seeing a brief angry facial expression within her explicit memory system denoting that there is a situation to be resolved. A moderate level of activation (a safe emergency) maximized the middle range of arousal and expanded the width of the therapeutic window.

Neurobiological Mechanisms of the Therapeutic Window

- Neuropeptide Y correlates with and balances cortisol level.
- Moderate activation of stress ensures that PFC circuits are engaged rather than retraumatizing.
- Modulates the adrenergic system and anxiolytic resourses to stay in the therapeutic window.

- A safe emergency acts on the social brain networks and accesses the parasympathetic nervous system.
- Moderating the excessive release of norepinephrine to minimize the risk of overshooting the therapeutic window.
- Extending the exposure for twenty minutes releases beta-endorphin to allow an analgesic buffer.

Michael's exposure to the implicit and explicit cues of driving a car again triggered a range of memories of the traumatic event, especially when approaching traffic. These exercises were repeated and extended (prolonged) so that the cues could reconsolidate as routine stimuli that he needed to habituate to so that he could commute to work without feeling overwhelmed with anxiety. The duration of the exercises was at least twenty minutes to allow for the release of beta-endorphin to buffer the anxiety. Because flooding could retraumatize him, titration within the window of tolerance started low and went slow. The principle goal was for the traumatic memories to lose their power. He was taught that the exposure process could be understood in the phrase "cells that fire out of sync lose their link," so that what was unnecessarily feared faded away. Of course, there was always the chance that another car would swerve into his lane or run a stoplight. The point is that he needed to adapt to a world of probabilities. Since driving at 2 a.m. on Sunday would increase the chances that drunk drivers were on the road, he drove defensively in the middle of the day when it was relatively safe.

The neurophysiological changes resulting from successful therapy:

- Increased size and activity of the dorsolateral PFC
- Increased size and activity of the hippocampus
- Decreased activity of the amygdala
- Moderated sympathetic nervous system activity when within the window of tolerance
- Decreased pro-inflammatory cytokines
- Recalibrated hypothalamic pituitary adrenal axis

Emotions as Cues

The emotions provoked during or after the trauma may later trigger a flashback. These conditioned emotional responses (CERs) can include fear, sadness, or horror (Briere, & Scott, 2015). For Michael fear and horror became associated with driving. Neutralizing the CERs required that they be activated, not reinforced, and then counterconditioned. Because the utility of perceiving, labeling, and accepting emotions can help diminish their intensity, therapy identified and modified the thoughts that exacerbated the CERs associated with the accident. Practical action through driving, in concert with purposeful thoughts about the necessity of getting to work, doing errands, and visiting friends, transformed into new meaning associated with those emotions. Therapy necessitated titrating the exposure to the CERs within the window of tolerance and at the middle of the arousal curve.

Neutralizing Conditioned Emotional Responses

Advise clients to do the following:

- Delay tension-reduction behaviors.
- Ride it out—understand that the feeling is only temporary.
- Hold off long enough to defuse the power of the feeling.
- Understand that the upsetting feeling will eventually become tolerable.
- Don't try to change the feeling, change your relationship to it.

Therapy took advantage of the fact that, every time a memory is retrieved, the underlying memory trace becomes once again fragile so that it goes through another period of consolidation. An incremental approach worked best so that the traumatic implicit memories were dealt with in smaller units and in a hierarchy from less distressing to more distressing. We broke down the levels of distress into doable chunks so that the highest-intensity emotion occurred in the middle of the session and then the session ended with calm. Novel and distinctive explicit memories were co-constructed in therapy to facilitate greater access a wide range of PFC and hippocampal circuits so Michael could apply affective top-down inhibition, so that the flashbacks diminished in intensity.

In sum, traumatic CERs periodically and abruptly trigger flashbacks. They need to be integrated within the explicit memory system to make them coherent and better structured and to reduce the risk of unwanted intrusions. When those traumatic memories are reactivated through exposure and reconsolidation they become less distressing.

AVOIDANCE, DETACHMENT, AND DISSOCIATION

Linda had typically "checked out" during Steve's abusive tirades. She said that the only way she could endure the beatings was to "go somewhere else." This method of protecting her "self" began during childhood when her drunken father terrorized everyone in the household.

Like other species, in addition to the sympathetic and neuroendocrine systems we possess a primitive defense system made possible by the unmyelinated vagus. During extreme trauma the metaphors of "deer in the headlights" and "playing possum" represent actual self-protective responses to extreme life-threatening threats. The freeze, collapse, or feigned death reactions play many roles in the wild. For humans, the spectrum of primitive defenses includes mild, moderate, and extreme detachment. For example, if her other means of defense did not work, Linda became detached and immobilized in response to horrific pain that she could not stop.

Detachment can occur at different levels. With mild detachment, or absorption, she may experience a breakdown in the ability to notice outside events such as how Steve was handling her in the moment. Her executive network checked out while her default-mode network dominated. Her distorted perception at times extended to an altered sense of herself, with the salience network checking out, too. During moderate detachment she may tend to have feelings of depersonalization and derealization and see herself as if from afar, as an observer. Finally, in extreme detachment she may become unresponsive and catatonic and have no sense of self or time.

The "continuum of deattachment" in disassociation (Allen, 2001) includes the following:

- Mild detachment: an altered sense of self and breakdown in the person's ability to notice outside events
- Moderate detachment: an experience of unreality, including feelings of depersonalization and derealization
- Extreme detachment: a state of unresponsiveness to the extreme of catatonia

Some people revert to these defenses more than others. Linda's propensity to quickly revert to these primitive defenses occurred in absence of a coherent salience network. As described in Chapters 1 and 2, "self"-organization is promoted by an internal sense of safety and positive attachment, which would have permitted her adequate self-awareness. But she endured adverse childhood experiences and abuse, and her attention was drawn outward, away from the development of a durable salience network. Her internal representations were fragmented in favor of being hypervigilant for potential danger. Later, as an adult, in response to extreme stress she had difficulty feeling her body and differentiating emotions. When she encountered acute stress or life-threatening trauma with Steve, she tended to detach from the situation.

Therapy entailed helping her build a coherent salience network monitored by a well-functioning executive network. Building the capacity for self-reflection within her default-mode network helped her develop a coherent internal life by helping her identify, label, and accept feelings. Because her schemas were primed by situations and feeling states associated with the trauma, they needed reconditioning through activation and reconsolidation. This process took time and necessitated working within the implicit memory system because her reaction patterns did not contain the contextual representations of the past or were not realistically relevant to the present.

The Somatic Therapies

When I presented a seminar to Afghan government service workers on coping with stress, during a break Aamir told me that when he says goodbye to his children in the morning he does not know if he is going to see them in the evening. Car bombs and IEDs engineered by the Taliban are quite common in Kabul. His anxiety centered not only on his family but also on his own safety—two car bombs had exploded near his office. He was painfully aware that working for the government made him a target. One day on the way home from work the car in front of him hit an IED, killing everyone inside.

His implicit memory systems encoded traumas as a mix of procedural and emotional memories that fused specific body movements and sensations with emotions. For example, when someone in the lecture hall accidentally dropped his briefcase, making a loud bang, Aamir abruptly ducked his head, only to raise it again with intense fear on his face. His automatic avoidant behaviors such as ducking became part of an array of procedural and emotional memories that represented somatic action patterns. They offered an important focus in therapy because they formed a key mechanism underlying trauma (Levine, 2015). Intense emotions combined with procedural memories that served as a guide to protect and defend him became chronic and corrosive to his sense of self. The goal was to bring him back to a state of dynamic equilibrium.

Jean Martin Charcot and his student Pierre Janet were the first to identify the implicit-nonconscious procedural memory system as holding traumatic memory and experienced through the body. Charcot, a neurologist working at the Salpêtrière Hospital in Paris the late 1880s, described jerky movements, contracted postures, and the tendency to suddenly collapse in response to cues to traumatic memory. Janet followed by describing trauma as replayed and reenacted through body movements and visceral sensations. These physical manifestations occurred as if the trauma were happening in the present, not as thoughts and reflections of what had occurred in the past. As raw sensations, they bring the past into the present, often with horrifying emotion.

Therapy that emphasizes body sensations attempts to address procedural and emotional memory circuits, which can be cued and expressed somatically. Implicit traumatic memories are expressed by the "somatic narrative" in gesture, posture, facial expressions, speech prosody, eye gaze, and movement (Ogden & Fisher, 2015). The body, indeed, "keeps the score" (van der Kolk, 2015).

Somatic based therapies, including the so-called bilateral-reprocessing therapies such as Eye Movement Desensitization Reprocessing (EMDR) and the tapping therapies such as Emotional Freedom Therapy, as well as Sensorimotor Therapy and Somatic Experiencing, all utilize somatosensory factors. Integrating the implicit and explicit systems through the somatic techniques can accelerate the narrative reorganization and strengthening of explicit memories.

It is important to find the common factors among these approaches and shed elements of the approaches that are superfluous (Arden, 2015). Along those lines, dismantling studies have not demonstrated that the bilateral movements affect symptom reduction (Cahill, Carrigan, & Frueh, 1999). However, the somatic-based therapies do disrupt traumatic memories to help clients reconsolidate and integrate them.

How does the body keep the score? Because much of the memory of the trauma is implicit, coded in body movements and sensations, access to them must be somatic. These procedural memories have been dysregulated from explicit memories and conscious awareness. One of the goals of therapy is to integrate them so that the procedural memories do not erupt without warning to trigger flashbacks.

Somatic-based approaches place attention on the physical sensations and support the cortical containment of traumatic memories into general semantic networks (Stickgold, 2002). Orienting identifies and shapes the amount and quality of the information processed that all species use to guide their behavior (Sokolov, Spinks, Naatanen, & Heikki, 2002). Redirecting attention to personally relevant information, such as a novel stimulus, promotes the orienting reflex.

The initial discovery of orienting response dates back to Pavlov (a few decades after the research on classical conditioning). He argued that his dogs "lost their sense of purpose" after being traumatized after a flood. In recovery, he argued that the dogs reoriented themselves to a sense of safety by noting that a previously frightening stimulus was not dangerous after all. He suggested that they reframed the new, non-dangerous stimulus by reorienting to a new identification of it, explained in the phrase, *Shto takoe?* (Что такое? or *What is it?*).

For humans, the reorienting response provides a much greater shift in brain activity. A stimulus that prompts a person to notice what happens next primes PFC activity. Coding in novelty, essentially an unexpected somatic sensation, integrates PFC, anterior cingular cortex, hippocampus, and basal ganglia circuits by moderate bursts of dopamine, and in so doing, orienting serves as a sort of a kickstart to the connectivity between the executive and the salience networks. Even the most subtle of stimuli can prompt a shift to the executive network.

Pierre Janet proposed that trauma victims are stuck and cannot accumulate new experiences. Somatic-based therapies offer a kickstart to reopen the gate. Consider how an unexpected stimulus such as cat darting across the road or an unexpected omission of a stimulus, such as when the hum of the refrigerator motor that suddenly turns off, can evoke the orienting reflex. Whereas we had habituated to the sound of the refrigerator motor, when it turns off it becomes novel. The orienting reflex is evoked when there is a discrepancy between a person's contextual expectations and a stimulus. Because one of the many functions of the hippocampus is as novelty detector, as well as a key part of the explicit memory system, novel somatic stimuli abruptly kindle the integration of the memory systems, which promotes affect-enhanced regulation.

Developing the flexibility of the top-down or bottom-up orienting and habituation responses helps us adapt to changing conditions. However, trauma victims may tend to dishabituate to a novel stimulus due to their hyper- or hypoarousal states. Whereas the bottom-up orienting reflex is reflexive, top-down orienting is cognitive. Being responsive to new information necessitates regaining a dynamic equilibrium in the attentional circuits and the mind's operating networks. Activating the orienting response in therapy promotes the ability to attentionally orient toward stimuli related to recovery, which is congruent with a hopeful narrative.

Somatic stimulation, whether induced by eye movements, bilateral stimulation, tapping, acupressure, or acupuncture, induces a shift in attention involving the orientation response (Sokolov, 1990). The orientating response may induce an REM-like state, which facilitates the cortical integration of traumatic memories when PFC and hippocampus activity has shifted (Stickgold, 2002). The reorienting of attention occurs automatically when a sudden movement grabs the client's attention, such as directed eye movements, taps, or intention movement of the body to specific postures. The reorienting of attention requires release of focus on one location or position so that attention can shift to a new location or position. Repeatedly reorienting attention from one location to another by somatic stimulation kindles the prefrontal and hippocampal networks to build here-now and top-down circuits of the executive network that put the past in the past. Though clients do not forget trauma, the power of the past becomes neutralized by the safe feelings coded in the salience network in the present moment.

In sum, somatic-type therapies evoke the orienting response to facilitate the integration of traumatic memories. Through focused therapeutic effort combined with somatic stimulation, these approaches activate the PFC and hippocampus. By combining exposure and somatic stimulation, the reconsolidation of implicit and explicit memory associated with the trauma reconsolidates by "safe emergencies" within the therapeutic alliance. Moderated degrees of anxiety within the window of tolerance facilitate the optimal conditions for therapeutic neuroplasticity.

POSTTRAUMATIC GROWTH

A major goal of therapy involves working with the client to develop a belief in a viable optimistic future that has meaning, to promote "posttraumatic growth." The potential for positive growth to occur posttrauma has been exemplified by reports that up to half of trauma victims describe some sort of positive outcome posttrauma (Updegraff and Taylor, 2000). Linda gained posttraumatic growth through her peer counseling efforts with other victims of domestic violence and adult children of alcoholics. By helping her develop a sense of meaning to move beyond the traumatic events, her posttraumatic growth involved constructive changes to her sense of self, her relationships, and her philosophy of life (Tedeshi & Calhoun, 1996).

Though the traumatic events were behind her, they are never forgotten. Her "self"-organization involved leaps to higher levels of insight and affect regulation as she acknowledged her interdependence with the world. The "self"-organizing process associated with posttraumatic growth transcended the pain of the traumas as she embraced greater meaning to her life. Her posttraumatic growth involved acknowledging the illusion of invulnerability without hopelessness by embracing realistic optimism about her life. This paradoxical recognition of vulnerability and hope promoted better control and a realistic sense of strength. The changes to her relationships allowed her to deepen intimacy and share feelings about what happened, as well as aspirations for the future. Through posttraumatic growth her old sense of self ("old me") transformed to a new sense of self ("new me"), with a recognition that there was no turning back to the old self. Her new sense of identity became wiser, and she used the opportunity to become more compassionate person. Her involvement as a support to others who had experienced trauma tapped into a natural inclination to gather with others for safety and healing. Increasing cognitive complexity and "self"-organization bolstered her resiliency as she expanded the number of different perspectives she had of herself to weather subsequent stressors.

The importance of developing meaning posttrauma was described in one of the most inspiring books in the mental health field, *Man's Search for Meaning* by Victor Frankl. Despite enduring the horrors of Auschwitz, Frankl embraced a transcending sense of meaning, demonstrating that deeply traumatized people can move on to wisdom and growth. In fact, many people who have been horribly traumatized have gone on to gain a deep sense of meaning and satisfaction with life. My Armenian relatives, despite living through genocide, thrived in the United States and France. Posttraumatic growth for my grandfather was operationalized by education. He spoke five languages and until a week before his death taught himself six new words a day. He instilled in me a quest for knowledge. On a political level posttraumatic growth can affect millions of other people as well, as personified by Nelson Mandela's expanding a sense of meaning and commitment to others.

NINE

Transcending Rigidity

> Start by doing what's necessary, then do what's possible, and suddenly you are doing the impossible.
>
> —Francis of Assisi

Scott and Annie fell into depression like it was a black hole. Previously, both had good-paying jobs, their two children were reasonably well adjusted, and they lived in a quiet, upper-middle-class neighborhood close to parks and shopping. When Annie lost her job she also lost her desire to do anything with the family. As her life constricted into rigid isolation, her whole family began to feel the gravitational force of the black hole. Scott became easily stressed and increasingly depressed. And with foul moods their children isolated themselves to their rooms with their computers. After a year of isolation from family and friends, they accepted a dinner invitation from their neighbors. When they discovered that one of them was a psychologist, the conversation quickly shifted to how miserable they felt. Annie complained of having no energy and feeling "stuck in a dark hole" all the time. Her doctor told her that she had chronic fatigue syndrome, then the diagnosis shifted to fibromyalgia, and then finally to prediabetes. She said, "They don't know what's up with me. I just feel ill and can't seem to get out of the hole." She turned to Scott, wagging her finger, "And this guy is no help!"

Scott shrugged his shoulders, hesitated, and then nodded his head in agreement. "Truth is, my job is torture. There are deadlines that I can't meet and a boss I can't please. The only time I have alone is during the two-hour bumper-to-bumper commute." Annie interrupted, "We have been on a boatload of antidepressants. Aren't you glad you've invited us?"

Both Annie and Scott became increasingly depressed, initially in response to stress and then isolation, poor self-maintenance, increasingly poor health, and the way they conceptualized their situation. This chapter explores the spectrum of depression and the multiple and interacting contributors. Because multiple factors contribute to depression, therapy must incorporate simultaneous interventions.

DIFFERENT SHADES OF BLUE

How common is depression, and is it a modern syndrome? Generally, one in ten people in the United States have suffered from depression once in their lives. Studies of current hunter-gatherer societies and extrapolating back in time to our ancestors indicate that depression was as common as among people in modern societies (Stieglitz et al., 2015). Despite its common occurrence throughout prehistory and recorded history, depression is not a discrete psychological disorder with one etiology and a clear set of symptoms (Maletic & Raison, 2017).

Women outnumber men with depression two to one. Women seek therapy for depression more often than men do, and when they do their symptoms appear on the surface like what is commonly regarded as depression. It was easier for Scott to talk about stress than depression. Indeed, he did suffer from a mix of stress and depression, but the stress symptoms were more obvious. Men may exhibit symptoms that can be misconstrued only as chronic stress, irritability, anger, and recklessness when depressed. Despite the differences, one common factor was that both Annie and Scott became ill, which increased their depression.

Gene-Environment Interactions

Because depression is not a finite disorder with one etiology and one recommended treatment, reductionism has led us down a blind alleys. The quest to find a gene that specifically causes depression has come up empty. Despite meta-analyses of genome-wide association studies involving thousands of subjects, no specific genes responsible for depression have been identified (Major Depressive Disorder Working Group, 2013). Genes, the environment, and individual behavioral differences interact.

As described in Chapter 3, one of the most widely researched gene-environment interactions associated with depression has been on the "short form" of the gene for the serotonin transporter. Compared to the long version, the short version is associated with an increased risk, not a cause, of developing mood disorders. The risk increases in response to stress, especially if the stress occurs in childhood (Sharpley, Palanisamy, Glyde, Dillingham, & Agnew, 2014). In people who suffered adverse childhood experiences (ACEs) or possess the short version of the serotonin transporter gene, the structures in the salience network could be undermined. On the other hand, social support can moderate the effects of the serotonin transporter gene, minimizing the risk of depression (Kilpatrick et al., 2007).

More women than men have the short version of the serotonin transporter gene, which is associated with anxiety and depression when switched on by threats or stress. The process is not as simple as once thought: the short version does not cause depression but, rather, tends to increase a person's sensitivity to the environment, whether stressful or positive (Homberg & Lesch, 2011). In fact, those with the short form and who also receive enhanced nurturing during childhood are more resilient later in life than those without the short form and had also received enhanced nurturning.

The interaction among ACEs, the short version of the serotonin transporter gene, the reduction of the affect regulatory neurocircuits, and the development of major depression can take many forms (Frodl et al., 2010). People who endured ACEs are at greater risk for developing depression and receptivity to treatment (Uher, 2011). ACEs combined with the short version and subsequent stress have been associated with an increased risk for suicide (Sharpley et al., 2014). Sociocultural factors can trigger epigenetic affects by the stress, inducing power disparity between men and women. For example, Annie crashed into the glass ceiling of nonequal pay for the same job. After being passed up for promotions, she took on more work delegated to her by the promoted men that she previously supervised. Eventually, she went on disability with a diagnosis of depression.

Neurotrophic Factors

One of the ways that our brain stays healthy is through the release of neurotrophic factors (meaning growth-enhancing substances). They enhance flexibility (neuroplasticity) and growth (neurogenesis). When these factors are not operative, our brain begins to shut down into rigid mood states we call depression and ultimately dementia. A variety of neurotrophic factors have been identified, including brain-derived neurotrophic factor (BDNF), glial-cell-derived neurotrophic factor (GDNF), fibroblast growth factor, and nerve growth factor.

BDNF is the most researched of the neurotrophic factors. It promotes brain growth, neuroplasticity, and neurogenesis, especially in the hippocampus. While it has been shown to enhance cognition and stabilize mood, its depletion or absence is associated with depression and dementia. Accordingly, people with low BDNF, such as those who are obese, tend to have smaller hippocampal areas and are also more inclined to suffer from depression.

Though BDNF has been getting all the press, GDNF has increasingly been shown to play a crucial role in neuronal and glial health, and so with cognitive functions, mood regulation, and resiliency to stress. It is an important modulator of monoamine synthesis and so plays a key role in the development and survival of dopamine, serotonin, and norepinephrine. These neurotransmitters are the main targets of many of the antidepressant medications. Each plays important roles in mood stability, energy, and motivation.

Genetics, Epigenetics, and Depression

The gene regulating BDNF synthesis has two different alleles, leading to different genotypes: People who carry the Met variant of the gene regulating BDNF synthesis and those carrying Val variant (Maletic & Raison, 2017).

- The Met/Met genotype is associated with decreased BDNF regulation and distribution; with structural brain changes associated with depression, memory deficits, and decline in reasoning skills; and with reduced gray matter volume in the dorsolateral prefrontal cortex, orbital frontal cortex, amygdala, and hippocampus.
- The Val/Val genotype is associated larger hippocampal volume and, if depression is present, a better response to treatment.
- Epigenetic changes, including hypermethylation of the BDNF gene, have been proposed to be a factor in depression and suicidality and may provide the platform for some of gene-environment interactions (Lockwood, Su, & Youssef, 2015).
- Epigenetic factors play a role with some women having a mutation in CREB-1 gene, which is turned on by estrogen and is associated with depression.
- Altered expression of GABA and glutamate transmitter genes has been found in depressed people who ended their lives with suicide (Fiori & Turecki, 2012).

A decline of BDNF and GDNF leads to demyelination. Diminished neurotrophic support alters energy supply and increases oxidative stress and neurotoxicity, all contributing to depression (Maletic & Raison, 2017). These corrosive dysregulations further contribute to depression by impairing the brain structures that regulate mood and executive function including the anterior cingulate cortex (ACC), the dorsolateral prefrontal cortex (PFC), orbitofrontal cortex, hippocampus, and amygdala.

Impairment in the ACC, a key part of the salience network, is associated with those depressed clients who have resigned, do not perceive conflict between the demands in the environment and their state, and have lost the will to change. With impairment to the dorsolateral PFC, a key part of the executive network, they perceive the conflict between their environment and state but are unable to activate goal-oriented behavior to facilitate change.

Inflammation-Depression

Many illnesses cause and exacerbate depression. Depression also makes many illnesses worse. An extensive study of approximately a quarter million people from sixty countries found that having a medical condition increased the risk of also having depression by 300–600 percent. The risk of depression increased to 800 percent with two medical conditions (Moussavi et al., 2007). Those illnesses associated with higher levels of inflammation are also associated with more depression than are illnesses with less inflammation (Raison & Miller, 2001). Inflammation appears to represent the central factor.

Both Annie and Scott suffered from the positive feedback loop between depression and their poor health. Scott was diagnosed with metabolic syndrome and Annie was warned that she was developing type 2 diabetes. The synergistic effect between depression and illness goes beyond the obvious that they were less motivated to engage in behaviors that manage those illnesses. With their energy dissipating, neither of them felt like doing the very things that could lift their depression.

Risk Factors for Depression Also Associated With Inflammation

- Medical illness
- Psychosocial stress
- Sedentary lifestyle
- Obesity
- Diabetes
- Metabolic syndrome

- Cardiovascular disease
- Infection
- Gut microbiome imbalance (dysbiosis)
- Autoimmune disorders
- Diminished sleep
- Social isolation
- Poor diet (e.g., skewed ratio of omega-3 to omega-6 and high in simple carbohydrates)
- Smoking and second- hand smoke
- Air pollution
- Winter for those with seasonal affective disorder

There are significant biological links among depression, inflammation, and ACEs. Early-life trauma or deprivation are associated with an elevated incidence of depression and illnesses such as diabetes and cardiovascular disease (Raison, Capuron, & Miller, 2006). Depressed people with a history of early-life trauma tend to have abnormal levels of the pro-inflammatory cytokines (PICs), such as interleukin-6 and tumor necrosis factor. These associations predict unfavorable courses of illnesses and treatment outcomes for depression (Nanni, Uher, & Danese, 2011). Inflammation represents a common factor between ACEs and adult poor health (Danese, Pariante, Caspi, Taylor, & Poulton, 2007).

Depression that runs in families tends to co-occur with various inflammatory illnesses. For example, some families that are positive for major depression are also positive for fibromyalgia and irritable bowel syndrome (Wojczynski, North, Pedersen, & Sullivan, 2007; Raphael, Janal, Nayak, Schwartz, & Gallagher, 2004). Factoring out second-hand smoke, children and adolescents of depressed parents are at an increased risk for asthma and other respiratory conditions (Goodwin, Wickramaratne, Nomura, & Weisman 2007).

Chronic inflammation can develop as a result of poor self-care practices, obesity, and/or metabolic syndrome and is associated with autoimmune disorders and cardiovascular disease, which are all associated with depression. These multidimensional causal interactions can impair social competence and contribute to interpersonal conflict, which combined with depression and hostility are associated with elevations in PICs. The relationship between depression and hostility can increase the risk of type 2 diabetes and cardiovascular illness, as inflammation increases insulin resistance (Suarez, Lewis, Krishnan, & Young, 2004). The more severe the depression, interpersonal conflicts, and hostility, the greater the mortality risk.

According to the Centers for Disease Control and Prevention, by 2009 suicide surpassed auto accidents as the number one cause of injury-related death in the United States. Though many factors contribute to suicide risk, it is important to note that as depression associated with inflammation increases, so does suicide risk. Plasma concentrations of the PICs (interleukin-6 and tumor necrosis factor alpha) are higher, while levels of the anti-inflammatory cytokine interleukin-2 are lower in those attempting suicide, compared to depressed people who did not attempt suicide or nondepressed people (Janelidzes, Matte, Westrin, Traskman-Bendz, & Brundin, 2011).

Overall, an association between depression and chronic inflammation has been a consistent finding in studies of people across the life cycle. Causal interactions occur among perceived stress, loneliness, depression, and inflammation (McDade, Hawkley, & Cacioppo, 2006). These interactions have gained so much interest that a syndrome referred to as "sickness behavior" has been consistently associated with depression.

Sickness Behavior

Since the mid-1990s, the range of symptoms associated with depression and chronic inflammation have been dubbed "sickness behavior" because those afflicted, like Annie, appear, feel, and behave as if they are ill. Sickness behavior is characterized by disturbances in mood, cognition, and neurovegetative behaviors that mimic major depression and contribute to it (Dantzer & Kelly, 2007). Chronic inflammation and sickness behavior associated with it are characterized by disruptions in appetite and sleep, as well as decreased social and self-care behaviors and deficits in learning and memory. When feeling overwhelmed with all these symptoms of sickness behavior, Annie felt ill, so she acted ill. Her sedentary behaviors inadvertently made her more depressed and feel even more ill. She thought that she needed to "recoup" or "get over" the "illness" by resting. When she made a brief attempt to pull out of the self-perpetuating spiral, she initially felt worse over the short term, so she disengaged before discovering that over the long term she would have gradually felt better had she stayed engaged. She escaped the brief period of discomfort, so she did not see the long-term benefit of engagement, that is, until she began therapy.

Symptoms of Sickness Behavior

- Feelings of helplessness
- Depressed mood
- Cognitive deficits
- Loss of social interest
- Fatigue
- Low libido
- Poor appetite
- Somnolence
- Pain sensitivity
- Anxiety
- Anhedonia

It is important to put in perspective how sickness behavior plays a role in adaptation. Considering that during our hunter-gatherer 50 percent of people died by adolescence of various types of infections and various pathogens, our immune functions and specifically short term inflammation became a prominent defense.

However, chronic inflammation adversely affects the central nervous system, resulting in lethargy and withdrawal. The effect of PICs on the basal ganglia slows movement and leads a person to lie low and conserve energy while the immune system fights off the threat. When ill, we hunker down and avoid movement, involvement with others, and engagement in long-term goals as we wait until our body fights off the virus or infection. In short, the identification of infection or a virus, facilitated by the activation of the body's inflammatory response system, is also a powerful depressogenic stimulus (Maletic & Raison, 2017).

There are major metabolic costs of mounting a fever. The activation of the immune system promotes sleepiness, especially slow-wave sleep and suppression of REM sleep. Also, when awake, slowed movements, social withdrawal, fatigue, and diminished appetite all conserve energy to fight off whatever ails us. The behavioral aspects of sickness behavior, such as irritability, anger, and even anxiety, could be understood as methods of keeping others away for the benefit of both the ill person and those immediately nearby.

Sickness behavior is essentially depressive behavior. When Annie and Scott failed to engage in regular physical activity and withdrew socially, they simultaneously potentiated depression and inflammation. Because they felt ill, less motivated, and less capable, they failed to exercise and to engage socially and experienced more depression. Like sinking in quicksand, they sank in more depression brought on by inflammation and then sickness behavior.

PICs contribute to depression by altering the levels of dopamine, norepinephrine, and serotonin through a variety of pathways that reduce the availability of their amino acid precursors, such as tryptophan, which is necessary for serotonin synthesis (Schiepers et al., 2005). Low levels of these neurotransmitters represent just one of the many inflammatory contributors to depression and anxiety. PICs also have significant effects on the hypothalamus and the hippocampus, which can play a role in sickness behavior.

PICs can activate the enzyme indoleamine 2,3-dioxygenase (IDO), which depletes tryptophan, the primary amino acid precursor of serotonin. Thus, IDO indirectly lowers serotonin. IDO catabolizes tryptophan into kynurenine and its metabolite, quinolinic acid. Elevated IDO and quinolinic acid have been associated with increased suicidality. Inflammation-induced quinolinic acid can spur excitotoxicity through direct activation of NMDA receptors.

Loss of neurons as well as glia cells in mood-relevant brain areas such as the subgenual ACC has emerged as one of the hallmarks of depression. Compromised functional and structural integrity has been found in the amygdala-ACC circuitry, along with reduced ACC, amygdala, and hippocampal volumes, which are associated with greater risk for depression (Pezawas et al., 2005).

PICs increase the activity in the default-mode network and the salience network. In so doing, Annie's physiological condition became her concern, rather than engaging in the immediate environment. As she succumbed to withdrawal and inactivity, her depression took a downward spiral. Less activity, lower energy to fuel movement, and fatigue led to feelings of anhedonia, so she had less motivation to maintain social ties. Her social withdrawal led to less pleasure, more time for rumination on the futility of making any effort, and feelings of worthlessness. Her therapy reversed this retreat. Increasing social support has long shown to buffer stress and help lift depression and has been shown to lower levels of inflammation.

It important to make clear to depressed clients that feeling ill makes them act ill and that if they behave like they are ill, the feelings of depression will increase.

Stress-Depression Synergy

Scott's depressive downward spiral began with stress. Changes to his neurotransmitters and neurohormones associated with stress led to progressively tenser and more depressing experiences. The commute to work and dealing with Annie's depression combined to prime his stress circuits. As Scott's depression increased, his stress toleration decreased. And as stress increased, so did depression.

Depression is associated with dysregulation of body's threat detection circuits, including the sympathetic branch of the autonomic nervous system, the hypothalamic-pituitary-adrenal axis, and the immune system. When out of balance, mood disorders, illness, and stress-induced release of epinephrine and norepinephrine combine to stimulate the release of PICs. Increased activation of his sympathetic nervous system and low activation of the parasympathetic branch increased inflammation and depression.

The link between anxiety and depression is associated with disruptions in the hypothalamic-pituitary-adrenal axis:

- Increased levels of cortisol, norepinephrine, and epinephrine (adrenalin)
- An enlarged and hyperactive amygdala
- Enlargement of the pituitary and adrenal glands, elevated corticotropin-releasing hormone, and adrenocorticotropic hormone hypersecretion

During his third therapy session Scott revealed that his mother was an alcoholic. In response to postpartum depression, she started to drink just after he was born. According to family members, she spent much of his first year in her bedroom, and he spent a great deal of time in the crib until his father came home from work. Studies that indicated that babies of depressed mothers have hypersecretion of corticotropin-releasing hormone (CRH) associated with maternal deprivation or severe stress. Their altered set point for CRH neurons and hypothalamic-pituitary-adrenal axis activity contributes to a tendency to overreact to stress during adulthood and develop depression.

Even short-term stress can prime depression. The levels of dopamine, norepinephrine, and serotonin can drop for ninety minutes after stress. Lower levels of all these neurotransmitters have been associated with depression. For example, low levels of dopamine are associated with psychomotor retardation and with decreased blood flow to the left dorsolateral PFC. Because the left PFC, relative to the right, can inhibit amygdala

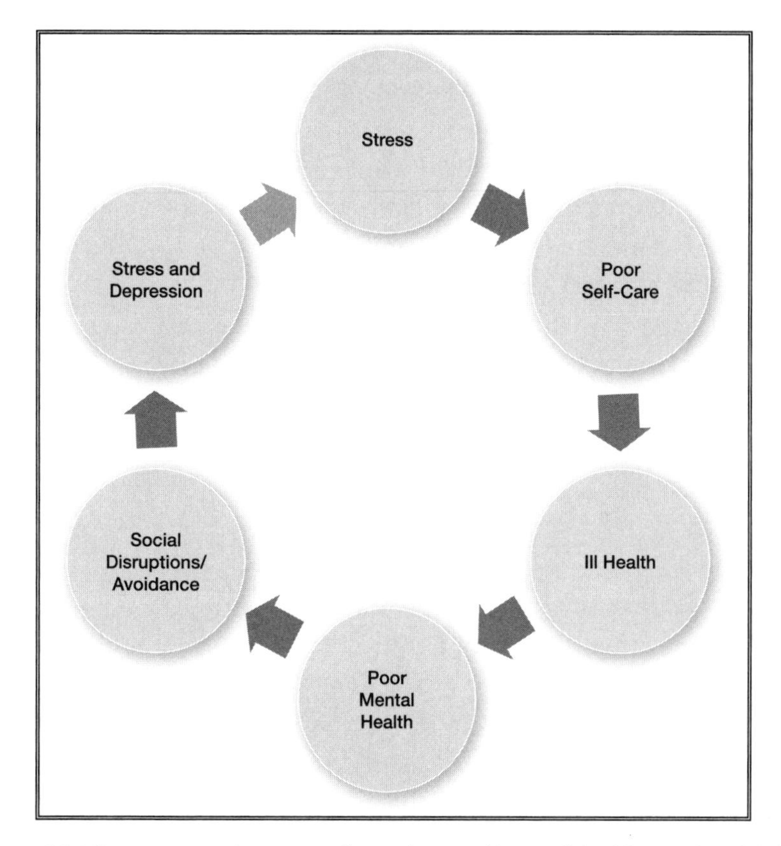

Figure 9.1: The synergistic depressive effects of poor self-care, ill health, social withdrawl, and stress.

activity, a depressed and anxious person like Scott needs to kindle the left PFC through approach behaviors.

The combination of depression and excessive amygdala activity promotes anxious and depressed memories (Davidson et al., 2003). Scott's emotional state when particular memories were encoded made it more likely to recall memories when he was that particular mood, referred to as state-based memory. In other words, when he was depressed and anxious, it was more likely that he recalled memories laced with both anxiety and depression. This did not mean that there weren't things that happened in his past that were positive. Essentially, state-based memories are the emotional lens that he used to recall memories.

The same applied to his perception of events in the present. His anxious and depressed emotional state tainted his perception, making neutral events feel negative. During periods of depression and anxiety this resulted in attaching emotional significance to normally trivial events and considering them important.

Scott developed a tendency to believe that the challenges at home and work meant either that he was cursed with bad luck or that God was testing him in some way. He wasn't quite sure which of these possibilities was the reason for his stress, but he believed that it was all happening as part of his fate.

Social Stress and Depression

Scott suffered from significant job stress in part because his stress level was high before he arrived at work. As he got ready to leave each morning, Annie told him how badly she felt and that he was partly at fault. During and afterward, he felt an agitated increase in depression. Then he endured the two-hour commute while ruminating on the depressing and anxiety-provoking events. He grew convinced that there was little in his past to feel good about.

Already stressed by the time he arrived at work, he tried to hide out from his boss and his coworkers. This combination of withdrawal and avoidance drew scrutiny and criticism from not only his boss but also his coworkers. His brain was primed for more threat, and he assumed that it would come from all of them. With his amygdala sensitized by cortisol, his anxiety became more likely during his periods of depression. Contributing to a downward spiral, an overactive amygdala led to elevated levels of norepinephrine, and CRH reduced his serotonin level, increasing the risk of depression as well as anxiety.

Suffering rejection that is difficult to reconcile can deflate self-esteem and self-respect. Worthlessness to others can promote social emotions such as shame and guilt. These socially related emotions represent conflicts with oneself and depression. The association of social stress, trauma, and depression has been shown across cultures and societies. For example, during a twelve-month period women in Zimbabwe had an 87 percent burden of stressful life events, and 30 percent reported simultaneous depression. At the other extreme, women in rural Basque reported no stressful life events and had a 2.5 percent prevalence rate for depression (Brown, 1998). The bottom line is that acute stress, especially if it involves humiliating, socially rejecting, entrapping feelings, is causally related to depression (Maletic & Raison, 2017).

Scott felt rejected, depressed, and ill at the same time. As noted earlier, social withdrawal protects others from our illness, and it is part of the constellation of symptoms of sickness behavior. Our capacity to form and maintain relationships has been fundamental to our evolution, development through the life cycle, and health. It is therefore no surprise that poor-quality or no relationships can have a devastating effect on mood. Just as being rejected, ostracized, abandoned, or isolated could have meant death to our hunter-gatherer ancestors—and suffering these major stressors as ACEs during development often results in multiple ill health consequences—so do they at any point during the life cycle result in anxiety and/or depression.

The ACC is involved in detecting social pain and physical pain (Muller-Pinzler, Krach, Kramer, & Paulus, 2016). The dorsal (top) portion of the ACC is sensitive to social rejection. Ill health can also sensitize the dorsal ACC, via PICs contributing to social anxiety. In other words, when ill we tend to feel rejected by others and depressed.

The ACC is positioned at the intersection of dopamine neurons, periaqueductal gray, hypothalamic, and amygdala circuits, making it crucial for the detection and regulation of arousal and drive. This makes the ACC important for instigating adaptive responses to negative emotional stimuli such as anxiety and depression. It serves as an important switch to either engage the fight-or-flight response or enhance the sensitivity between the salience and executive networks to bodily distress signals (Maleic & Raison, 2017).

Revolving Internal Conflict

Both Annie and Scott had been feeling paralyzed by indecision, not knowing what to do to get out of their depressive rut. They felt like their circuits were jammed. Accordingly, depressed people tend to not activate the ACC or to shut it down in the face of difficult situations. The ACC is a monitoring system that activates when we face uncertain situations or demands that may have multiple interpretations, some of which potentially could go wrong. Active during conflicts, ambiguous situations, and consciously experienced feelings, it is important to help negotiate difficult situations. The ACC is chronically underactive among depressed people, and its activity increases when depression lifts.

The cognitive and affect regulatory deficits with depression are associated with reductions in the following:

- Neuronal densities in the dorsolateral PFC
- The ventral PFC
- The ACC

Abnormal ACC functioning has been one of the central neuropathic correlates of depression. The dorsal ACC has generous connections with the dorsolateral PFC, and so executive network functions such as working memory and decision making suffers. In contrast, the rostral/ventral (bottom) part of the ACC is involved in emotions and is preferentially connected to the amygdala and so with threat detection (Davison et al., 2002). Sadness overactivates the ventral ACC, the amygdala, and the insula, reducing of activity in the dorsal ACC and dorsolateral PFC.

Depressed people tend to get lost in negative ruminations, brooding, gloomy autobiographical memories, and depressive feeling states, leading to hopelessness. These tendencies reflect reductions in the dorsolateral

PFC and ACC that strengthen connections between the salience network and the default-mode network at the expense of the executive network (Maletic & Raison, 2017).

The Depression Switch

Seriously depressed people tend to have excessive activity in the ventral ACC. Increased metabolism, increased blood flow, and reduced volume of the ventral ACC have been associated with depression, in step with the intensity of sadness (Mayberg et al., 1999). The area adjacent to the ventral ACC (referred to as Brodmann area 25) has been referred to as a "depression switch" because of its bridge between attention and emotion. The improvement of depression is associated with a reduction in activity in the ventral ACC, insula, and orbitofrontal cortex, with corresponding increases in the dorsolateral PFC, hippocampus, and dorsal ACC (Mayberg et al., 2005; Goldapple et al., 2004). When depression is successfully treated, these areas tend to show increases in hippocampus and ACC, with a concommital decrease in rumination and increase in optimism (Sharot, Riccardi, Raio, & Phelps, 2007).

Our ability to detect and correct errors is essential for our mental health. The ACC is critical for error correction and to cognitively resolve errors. Cognitive reappraisal involves recognizing the negative pattern your thoughts have fallen into, and changing that pattern to one that is more effective. Detecting errors is generally excessive for people with depression, and Scott's therapy addressed these overly critical tendencies so that he was better able to make decisions and engage in behaviors without second-guessing himself. Successful reappraisal is associated with increased activation of the ventral lateral PFC and the nucleus accumbens (Wager, Davidson, Hughes, Lindquist, & Ochsner, 2008). However, a one-dimensional reappraisal has limited value and durability, such that reappraising his situation with the narrative that he "will hold together until Annie recovers" was tested quickly and failed. By co-constructing a broader range of meaning and feeling, his reappraisals address his stress and depression head on instead of with his usual avoidance.

Approach Versus Avoidance

The secret of getting ahead is getting started.

—Mark Twain

Scott's increasing avoidance of peers at work and his family had been painting him into a depressive and anxious corner. While avoidant and withdrawal behaviors are associated with the overactivation of the right hemisphere, approach behaviors are associated with the left hemisphere. Accordingly, depression is associated with a deactivation of the left PFC and the motivation circuit. The most vivid example of the effect of this asymmetry is illustrated by patients with a left-side stroke, making the right hemisphere dominant. They become extremely depressed and may wonder why they should make the effort to go on living. In contrast, those with a right-side stroke, which makes the left hemisphere dominant, show a laissez-faire attitude, demonstrating more acceptance of their condition, and may regard the stroke as merely an inconvenience. The bottom line is that avoidance and withdrawal contribute to depression and that engaging in life with constructive behavior diminishes depression.

> Regions of each side of the PFC contribute differently to behavior and affect: the right dorsolateral PFC is associated with avoidance/withdrawal behaviors, while the left dorsolateral PFC is associated with approach behaviors, and the right orbitofrontal is associated with negative emotions, while the left orbitofrontal cortex is associated with positive emotions.

Because Scott tended to be overwhelmed by a global perspective and withdrew from others, he overactivated his right hemisphere, which resulted in more negative emotions. To rebalance the ratio of activity between the two hemispheres, I encouraged him to construct short-term goals that were described with a positive narrative and to engage in accomplishing these goals. All these promoted a "can do" attitude and positive emotions, boosting activity in his left hemisphere.

Because it is much easier to agree that taking action is necessary than actually doing so, therapy with Annie and Scott needed to address their motivation. To put the importance of behavioral effort in perspective, I have often used the analogy of starting a car that has lost its starter. To start my semi-electric Volt, I merely press the start button. But if the starter is out, the button does nothing. Similarly, having an insightful thought is not enough to start up motivation to lift out of depression. The biological factors contributing to depression are too strong to be reversed by thought alone. I once owned a 1967 VW squareback that had a manual transmission, and on a trip to the Rockies in Colorado my starter broke. I was able to start the car anyway by parking on a hill or having friends push me while I sat in the car with my foot on the clutch peddle and the transmission "in gear." After reaching 5 miles per hour I popped the clutch and the car started. Climbing out of depression, especially for Annie, necessitated this kind of approach. No insightful thought was enough to kindle motivation. Her motivational circuit needed to be primed much like my VW needed to be pushed. Because this circuit is essentially the habit circuit, motivation needs a push start to develop the neuroplasticity to create positive and mood-enhancing habits.

As described in Chapter 6, the nucleus accumbens, striatum, and the amygdala are in close proximity, and together they play a significant role in the motivational system. The nucleus accumbens is a peanut-size structure positioned deep within the salience network, between the striatum and amygdala. Because the striatum is involved in habits and movement, while the amygdala is involved in relevance detection, this habit circuit provides an interface between pleasure, emotions, and our actions. With executive network connections within the dorsolateral PFC controlling problem solving and planning, kindling motivation combines decision making with movement and follow-through.

Prior to therapy Annie failed to engage in rewarding activities; consequently, there was less activity in her nucleus accumbens, which resulted in a corresponding loss of pleasure. She became anhedonic, in part because her accumbens was deactivated. When she failed to move her body, there was a drop in activity in her striatum, which resulted in sluggish movement. And when she was less attentive to the here-and-now and engaged in less planning for the future, there was a drop in the activity in her PFC, resulting in poor concentration, which she described as "brain fog."

The effort to reestablish pleasurable activities has been a staple therapeutic approach for depression for over forty years, referred to as the behavior activation technique within cognitive behavioral therapy. Annie was encouraged to engage in pleasant activities. By monitoring her mood during these activities, she grasped the connection between behavior, cognition, and emotion.

Annie's motivation was primed by the interactions between her nucleus accumbens and incoming sensory information to evaluate emotional memory by her PFC as to whether to "go" with a healthy behavior. When she felt that a behavior resulted in a positive outcome, she strengthened those synapses mediated by dopamine, and her motivation increased to engage in that behavior again.

This neural circuit, also referred to as the effort-driven reward circuit, underlies why the behavior activation technique helps lift depression (Lambert, 2008). When the nucleus accumbens, the striatum, and the PFC were simultaneously activated, this kindles a positive mood. When the nucleus accumbens was activated by dopamine, Annie anticipated pleasure and was more motivated try to replicate the behavior again. The striatum-driven movement toward the positive habit and her PFC-generated executive functions of planning orchestrated her motivation to continue.

The motivation circuit connected her movement, emotion, and thinking, which helped lift her from depression. Because kindling these neural circuits necessitated use and follow-through, engaging incrementally in productive activities raised the levels of dopamine and serotonin, resulting in increased positive feelings, and allowed her to reap rewards of her productive behavior, boosting her sense of self-control, and self-esteem.

> It is common that depressed clients do not know where to start. They can be informed that they can rev up their brain circuits to lift depression by doing the things they don't feel like doing. Just by planning activities they kindle the PFC. Then when they move to perform the activity the striatum activates. And finally the nucleus accumbens activates when they feels the anticipation of pleasure of doing it again. Activated all together they can lift themselves out of depression.

REBALANCING THE MENTAL NETWORKS

Bob complained that his wife and friends suggested that he needed to be evaluated for attention deficit disorder. He disagreed with their opinion but did acknowledge that "at times that I'm lost in thought and get a little down." He added, "I just can't get some things out of my mind." Through his fifteen years in the Alcoholic Anonymous program he understood that he was "doing a lot of stinking thinking" related to events that took place when he was drinking. However, he argued that he was trying to come to terms with hurting and being hurt by his ex-wife.

Bob's pattern of default-mode network (DMN) activity promoted depression. His "reflections" became increasingly laced with feelings of threat, anxiety, and hopelessness. Without participation of his executive network, these ruminations took on a life of their own, with less grounding in practical reality. Bob excessively drudged up self-recriminating memories with regret, remorse, and self-pity. It was not productive time, to say the least.

> Hyperconnectivity of the DMN has been a consistent feature found in imagery studies that underlie the common symptoms of depression, including negative ruminations, brooding, and excessive self-referential thought. This hyperconnectivity is so powerful that it is evident in people who have recovered from major depression and may underlie their vulnerability to relapse (Nixon et al., 2014).

Responding less to external cues, Bob was rudderless and lost in a morose sea. With tendency to be "stuck" in himself at the expense of adapting to his environment, his depression was characterized by excessive coupling of the salience network with the DMN, producing ruminations on negative autobiographical memories and negative emotional states. His ruminative stewing hijacked his working memory and other executive network functions. The emotions generated from his salience network transformed into sadness, promoting the access of state-based depressive memories.

> ***Somatic Grounding***
> People with major depression may tend to have less insula activity. In fact, insula hypoactivity has been considered a risk factor for depression (Liu et al., 2010). This suggests that those who are less in contact with their body tend to be more vulnerable to depression. In contrast, increased activity and volume of the insula are associated with the successful treatment of depression (Fitzgerald, Laird, Maller, & Daskalakis, 2008). Bob regarded himself as uncoordinated and "terrible at sports" as a child. Therapy entailed increasing his attention to somatic sensations.

Bob's failure to flexibly engage and get his mental operating systems to work together disrupted the dynamic equilibrium between them and his mental health. When his salience network and DMN became coupled together, they minimized activity in his executive network, and he became lost in melancholy rumination and depressive emotions. He ruminated on what he had endured and felt more depressed because of it.

With depression there tends to be a short-circuiting between the salience network and the DMN that allows negatively valenced emotions to enter a reverberating ruminative loop (Maletic & Raison, 2017). This loop combines state-based negative memories of past failings and disappointments with depressive feeling states, brooding, and sulking. The cognitive blunting is more like smog than fog, breathing it in as toxic. Whereas excessive DMN activity is associated with depression, an increase in the executive network activity in problem solving helps navigate out of this depressive state.

Bob's DMN increased when his dorsolateral PFC was not engaged, such as when he was bored, experienced no novelty, or was just tired. His obsessive ruminations took over, and he stewed over the negative experiences he had with his ex-wife. Helping him break out of his depressive ruminations involved strengthening his attention skills to orient to the present moment, to reflect on what was occurring around him.

Because breaking out of a depressive DMN spiral was difficult, Bob was encouraged to work toward making the ruminations fade with exercise, social activities, and mindfulness, all demanding a here-and-now focus. For example, he engaged his executive network when he was jogging, briskly walking in the park, or having a conversation— during these activities he was less likely to drift into his depressive ruminations.

Therapy brought balance to the mental operating networks. His salience network provided information about important changes about his body sensations in response to what was happening in the environment and encoded the flow of his emotions. His DMN helped him reflect on himself and others, to access autobiographical memories, and to plan for the future and fantasize about possible positive outcomes. And his executive network took the information into working memory, prioritized, problem solved, and maintained attention, and he followed through on practical goals, behaving adaptively.

Between the Extremes

> Most people are about as happy as they make up their minds to be.
>
> —Abraham Lincoln

Abraham Lincoln suffered from periodic depression before and during the Civil War. Though the war took a major toll on him, simultaneously he suffered the death of his son and the depression of his wife, Mary Todd Lincoln. Despite the effects of acute stress and depression, he was able to manage the epic national crisis, support Mary, and orchestrate the deliberation of the "Team of Rivals" in his cabinet. Perhaps it was his capacity to reconcile opposites that enabled him to transcend depression (O'Hanlan, 2017). Being able to integrate and appreciate what is between the opposites offers a method of escaping the rigidity that depression brings.

Lincoln's ability to invite and then reconcile opposites offers an opportunity to transcend depression. Bob developed the capacity to hold in mind the shades of gray between the memory of the pain he caused in the past and his desire and effort to be a better person in the future. He transcended the black-and-white thinking, overgeneralizing, and catastrophizing that had contributed to his depression. The cognitive skills to cultivate and appreciate the shades of gray and ambiguity strengthened his resiliency and diminished his depressive ruminations in his DMN.

> Depressed people who incur damage to their hippocampus tend to lose the ability to appreciate the shades of gray and complexity inherent in most situations by reacting in a black-and-white manner and overgeneralizing. These deficits are based on faulty "orthogonalization" of information that would normally ensure that encoding new patterns does not interfere with the old and remains separable. We can say that Lincoln was a master "orthogonalizer." Normally, sufficient numbers of dentate gyrus neurons, made possible through neurogenesis, facilitate orthogonalization.
>
> A significant percentage of chronically depressed people have explicit memory deficits. And many people with chronic depression and those who were previously depressed show hippocampal shrinkage in the range of 8–19 percent. Potential causes of the shrinkage include increases in cortisol; excess glutamate that saturates the NMDA receptors, resulting in excitoxicity; and blocked neurogenesis, which results in no regeneration of hippocampus.

Effective psychotherapeutic interventions improved the connections between Bob's salience and the executive networks (Maleic & Raison, 2017). Meanwhile, meditation weakened the connections between his salience network and DMN while strengthening his executive network. Bob practiced mindfulness, maintaining nonjudgmental attention to the present moment, pulling himself out of the past, and noticing the subtleties in present. He worked on neutralizing ruminations through observation of what he was doing in the present moment. When he found himself drifting into ruminations, he brought himself back to the present moment to reflect on goal-directed behaviors. He discovered that he could not make up for the past but understood that he was responsible for the present.

As he practiced mindfulness, Bob attempted to maximize attention to the novelty in each moment. Given that his depression was kindled by his focus on the past, his here-and-now observations provided an antidote to his negativistic rumination. Mindfulness targeted his depression by neutralizing monotony with attention to subtle changes in his environment and cultivation of curiosity. Through affective labeling, he learned to say, "That's a depressing thought," rather than "That's depressing." As he built the capacity to decenter negativistic thoughts and feelings, he understood that momentary events were not permanent realities. Because he had often been on autopilot, teaching him to behave with intentionality helped establish here-and-now focus to break out of depressive automatic thoughts and behaviors.

Orchestrating a Broad Approach

Since immigrating to the United States, Lyudmila sought therapy in hopes of alleviating depression. She complained that her therapist had suggested interventions that she had already tried with her previous therapists. To each therapist she said, "I tried that and it didn't work." In fact, she had seen many therapists over the previous few years. None of them seemed helpful enough to get her out of the dark, overwhelming hole.

Because multiple factors can contribute to depression, therapy for Lyudmila necessitated a concerted effort of multiple and simultaneous interventions. Much like a conductor of a symphony orchestra, the integrated approach involved the simultaneous orchestration of multiple feedback loops. Because depression can result from multiple factors, it was imperative to apply simultaneous therapeutic interventions.

Promoting the functions of the left PFC by encouraging active behaviors, instead of the avoidant and withdrawal tendencies of the right PFC, helped balance out the set point, the relative ratio activation between the two hemispheres. Because negativistic and self-deprecating rumination was often part of her cognitive set, helping her switch off of the "depression switch" in the ACC by challenging her tendency toward self-criticism was an ongoing process.

Because many factors can contribute to depression, she needed to understand that she had to do all the things we talked about doing at once, to climb out of depression. Tempering the right PFC bias by activating the left PFC was achieved by encouraging her to engage in specific detail-oriented behaviors.

Interventions That Bolster Underactive Brain Areas

- Social engagement
- Practice aerobic exercise
- Attend to sleep hygiene
- Improve the diet, including omega-3 intake
- Use inquiry to counter mood-congruent bias in the hippocampus
- Rebalance the left/right PFC with details and activity
- Engage the motivation circuit with goal-directed behavior
- Use mindfulness to neutralize ruminations and monotony

I began this chapter by noting that depression does not represent a finite disorder related to a specific gene and etiology. Multiple feedback loops contribute to the array of symptoms we refer to as depression. So, too, do we need to address depression with a wide spectrum of simultaneous interventions.

TEN

Mind in Time

> If you had faith even as small as a mustard seed, you could say to this mountain, "Move from here to there," and it would move. Nothing would be impossible.
>
> —Jesus

Cara wanted quick relief from feeling depressed, and she found it. She was so pleased with her primary care physician for putting her on an antidepressant that she called his nurse two days later to report that she felt "100 percent better!" The nurse sounded perplexed instead of pleased. Linda sighed and then asked, "What's up? Are you so burnt out that you can't feel good that your patients are getting better?"

The nurse responded, "Ah . . . of course . . . I'm happy to hear it." But she still sounded vague. With a huff, Linda demanded to talk to her doctor. He soon called her to follow up. Linda asked, "Don't you and your nurse want to know that I am feeling better?"

"Well, uh, she, uh," he began. "Well, actually those medications generally don't work until three to four weeks. But, we are so pleased that you . . . "

"What do you mean?" Cara cut in. She grew quiet. Feeling immediately glum, her depression was back like an avalanche.

BELIEF, PLACEBOS, AND THE MIND

Expectations have a powerful effect on the outcome health care treatments. The placebo and nocebo effects represent the power of beliefs that a patient like Cara forms that she will either get better or get worse as a result of a particular intervention or medication. These effects have been studied extensively in medicine during the last forty years. Theoretical support for placebo effects have been an integral part of various schools of psychotherapy, especially hypnotherapy, psychosomatic medicine, and health psychology.

A placebo can be any therapy causing psychophysiological effects or used deliberately in a clinical situation, such as when the doctor knows that there is nothing pharmacologically active in the pills but hopes the patient will do better anyway. Or placebos can be included in treatment unintentionally when the doctor himself believes that the pills will be effective but is not aware that there is nothing of benefit in the pills. Most relevant to psychotherapy is that the placebo effect illustrates how beliefs can alleviate suffering and even change biology.

The early history of medicine may partly be a history of faith in a particular treatment resulting in placebo effects. Such treatments as lizard blood, crushed spiders, bear fat, fox lungs, blistering, plastering, leeches, and bleeding may have periodically had benefit through the placebo effect. While some herbs bought today at the local health food store do actually have some medicinal benefits, others have merely a placebo effect. The key is the patient's faith, belief, and hope in the treatment.

Cara had not only expectation of benefit but also faith that her doctor had the competence to provide relief from her suffering. Her belief in the treatment was augmented by feeling cared for beyond what family and friends could provide. The combination of empathy and expertise generated faith in his treatment. But when he informed her that the medication would not have an effect for weeks, her expectations abruptly changed, and so did her mood. Had the nurse initially responded by saying, "We are so pleased!," Cara's expectations and mood probably would have been quite different.

The following characteristics of a health care provider tend to increase the placebo effect:

- Good listening skills
- Empathetic attention
- Gaze attunement
- Appropriate touch
- Matching communication style (language and prosody)
- Welcoming physical appearance
- Close physical proximity
- Asymmetrical power dynamics between health care provider and patient (based on Kradin, 2008)

Incidence of a placebo response ranges from 10 percent to 70 percent, with effect averages of 35 percent (Kradin, 2008). The evidence suggests that placebos work best for subjective outcomes like pain. Placebos can be about half as effective as injection of morphine for pain reduction. When people feel that a particular medical intervention will hurt them less if pain relievers are applied, they will generally report less pain. For example, researchers applied a fake analgesic cream to two groups of people. Both groups had their forearms heated to a painful level. One group was told that they received an analgesic cream. The other group was told no medication was in it. The group that was told they had the medicated cream reported relatively less pain (Wager et al., 2004). There is also some evidence for efficacy in objective conditions (ulcers, angina), and positive placebo effects have been shown for insomnia, pain, and depression (Benedetti, 2008).

Cara experienced a type of placebo effect that reveals much about the major shift in mental health treatment. Over thirty years ago cracks began to show up in the foundation of the medical model in psychiatry, in part, because the placebo effect was revealed in a variety of studies for the efficacy of antidepressant medications. For example, a meta-analysis of three thousand patients who received antidepressant medication, psychotherapy, a placebo, or no treatment found that 27 percent of the therapeutic response was attributed to drug activities and 50 percent to the psychological factors surrounding the administration of the drug, which were unrelated to pharmacological activity (Sapirstein & Kirsh, 1988).

The placebo effect is indeed partly "in the head," as revealed by changes in brain activity apparent in several imaging studies. We may speculate that Cara's placebo response was consistent with responses of patients who received a placebo and were observed in brain imaging studies to show changes similar to those of patients who received Prozac (Mayberg et al., 2002). The placebo group, like the Prozac group, activated their prefrontal cortex, posterior insula, and posterior cingulate, with decreased activity in their hypothalamus, thalamus, and parahippocampus.

Studies have shown that people who believe that they are receiving a pain killer release endogenous opioids in their brains and, accordingly, report less pain. As an illustration of the role of endogenous opioids in the placebo effect, when they received opioid-blocking drugs, such as naloxone, they no longer reported pain reduction. Expectation also plays a major role in the placebo effect: when people are told that they are receiving a pain killer, dopamine is released in the nucleus accumbens as they anticipate relief from pain (Scott et al., 2007).

Placebo-Nocebo: Two Sides of Belief

The flip side of the placebo effect is the so-called nocebo effect: the negative effects from a placebo. Upward of 25 percent of patients receiving a placebo report adverse side effects, with 7 percent reporting headaches, 5 percent somnolence, 4 percent weakness, and 1 percent nausea and dizziness (Kradin, 2008). Whether positive or negative, the placebo response can be understood as a dramatic shift in the mind-body. For a nocebo response, mind-body systems self-organize away from an attractor governing homeostatic normality, while for a placebo response the movement is back to its previously established attractor of homeostatic normality (Kradin, 2008). The term *attractor*, borrowed from nonlinear dynamical systems, is a property of complex adaptive systems to explain how new order emerges from chaotic conditions. Some have suggested that we rename the placebo response the "meaning response" (Moerman, 2002). The meaning of a placebo or nocebo can be considered the attractor.

> Like positive placebo effects, nocebo effects have also been identified with specific brain areas associated with pain, including the anterior cingulate cortex and insula, key parts of the salience network. Our anticipation of pain—in other words, expecting to receive a painful stimulus and what to do about it—plays a role in the perception of the pain by the prefrontal cortex through its gating and filtering of sensory information (Koyama, McHaffie, Laurienti, & Coghill, 2005).

Because expectation frames perception, the belief in psychotherapy can have major effects on outcome. In Ericksonian hypnosis the "expectancy set" had been understood as a powerful ingredient for positive outcome. Forty years ago, while giving a grand rounds talk at the University of New Mexico, Jerome Frank, psychotherapy researcher and author of the *Power of Persuasion*, hypothesized that one of the most powerful variables to psychotherapy benefit was the client's expectation of relief of suffering. I looked around the room at the psychiatric residents as they scratched their heads. One of them chirped up incredulously, "If what we are doing can be boiled down to influencing the patient's belief that he will get better, then what's the point of our training in psychotherapy techniques?" To that Frank promptly replied, "You're here to gain confidence and then exude it to your patients." He went on to say that psychotherapy is garnered with many rituals and socially agreed upon expectations that something special will happen. To drive home that point, our offices are anointed with degrees hanging on the wall, sessions last fifty minutes, and as therapists we exude empathetic confidence. All these contextual specifics and meaning-making distinctions ec effect of the particular therapeutic technique employed. Had Cara been embedded in a society that believed that prayer, sand painting, or ritual dance were curative, her faith in one of those interventions would have influenced her recovery.

Yet, belief alone is not enough. Top-down and bottom-up processes interact. If Cara believed that she was receiving a medication that would not work until three to four weeks, she may have endured that much time before responding positively. Though her belief influenced the efficacy of the medication, it did not completely determine the outcome. She may have experienced some, a lot, or no relief, influenced not only by her belief but also a variety of other mind-brain-gene feedback loops, as described throughout this book.

The placebo effect demonstrates how the mind self-organizes felt meaning down to biological levels. Not merely a top-down causation, the placebo also represents constructed meaning emergent from bottom-up sensations. After Cara's conversations with her nurse and doctor, she began to notice a variety of side effects from the antidepressant, of which she had been previously warned. Because the sensations were called *side effects*, she believed that the medication must be slowly working. In response, she felt less depressed. Was this reaction, too, a placebo effect? It may have been a combination of both placebo and medication.

The point is that placebo effects work through interactions with multiple social and biological factors. Top-down belief and bottom-up biological processes interact to contribute to outcome. Belief alone does not determine mental health, but it is an indispensable factor.

> The greater the number of reasons to believe in the effect of psychotherapy, the greater likelihood that the client will experience a positive outcome. In other words, the more the client believes that there are biopsychosocial factors, including genetic, immune, brain activity, and cultural, that provide evidence-based scientific support for the interventions, the better the outcome. While modifying one belief factor may produce a ripple effect, changing others, the shear critical mass of multiple factors combined produces long-lasting mental health.

Attitude and Positive Psychology

> I am fundamentally an optimist. Part of being optimistic is keeping one's head pointed toward the sun, one's feet moving forward. There were many dark moments when my faith in humanity was sorely tested, but I would not and could not give myself up to despair.
>
> —Nelson Mandela

Taylor moved to the Bay Area, enticed by a lucrative job offer to join one of the tech giants of Silicon Valley. As a software engineer, he perfected algorithms that no one else had, making him a valuable commodity in an intensely competitive field. Initially, the job brought him a great deal of satisfaction and pride, buoyed by praise from management and peers. He found himself trying to fit in with his peers, emulating them by buying a $100,000 car, eating in high-end restaurants, and consuming only the finest wines.

To his dismay, the glow of the new job and environment soon faded. He thought his success would impress his father, but like everything else he had accomplished, it was never enough in his father's eyes. He and his wife bought a house high above Los Gatos with a spectacular view of the bay on one side and redwood forest on the other. That did not impress his father, either.

The novelty was gone as he established what he called a "monotonous" routine at work. He became convinced that his vacations to Paris, Bali, and the Seychelles Islands, all to five-star hotels, were essential for "downtime." But even the luxury and pampering began to wear thin. He expected the same level of extreme gratification from everything and everybody. His dopamine pathways had become tired. Prompted by what he considered a hollowness to his life, he shifted to wanting without liking and was left feeling empty yet wanting more.

Having habituated to an exceedingly comfortable lifestyle, a stable nuclear family without challenges, and a job that eventually demanded little of him but paid enormously well, he took for granted his life situation and wanted more. More of what, he wasn't sure, but it seemed like "something was missing." Dissatisfied with emptiness, he began seeing a therapist to discover the "right diagnosis." Had he seen a therapist embracing the archaic medical model, the diagnosis might have been depression, followed by medication. But he was fortunate to see one of our postdoctoral psychology residents, Sara, who was well schooled in the neuroscience of the pleasure pathways and combining it with research in positive psychology. She was especially familiar with the research on life satisfaction, known in the popular press as happiness.

Positive Psychology Has Deep Theoretical Roots

The paradox of happiness was described by Taoist philosopher Chang Tzu (350 BC), who said, "You will never find happiness until you stop looking for it." More recently, the fleeting and unexpected moments of life satisfaction have been described as "stumbling on happiness" (Gilbert, 2007). What is new is that research has been applied to the theory. Attitudes have been associated with different emotional outcomes.

To Taylor's initial dismay, Sara suggested that he work on cultivating the capacity to practice gratitude. He told her that she must be "stuck, just like everyone else, on the idea that wealth and happiness were the same thing. How trite!" But the so-called global happiness reports have consistently shown that wealth alone does not buy happiness. In fact, lotto winners generally gravitate back to their earlier mode of functioning after the initial glow of their new-found riches wears off. Like a Lotto winner, Taylor found himself back to self-criticism, the echo from his father, right where he started prior to taking the lucrative new job. Like a drug, his lavish lifestyle exhausted his sense of satisfaction. He felt as if his chance of happiness was flushed down Donald Trump's gold-plated toilet.

Researched Positive Attitudes and Behaviors

- Gratitude
- Compassion

- Acceptance
- Optimism
- Forgiveness

Taylor maintained that the real issue for him was that his father's message that he was never good enough meant that "enough is never enough." The truth was that his father did not feel good enough. In short, the misery was his father's emptiness, which didn't have to be shared with Taylor. Feeling less responsible for his father's feelings took time. One session was canceled when his wife had the flu. She was not completely recovered when he returned. He complained that without her usual expression of warmth and playfulness he felt at a loss. Certainly he was a beneficiary but also he enjoyed seeing her contentment. He said that she had the ability to be satisfied and grateful wherever they were.

At the most basic level, the pivot toward gratitude began for him as a simple expression of the thankfulness for her recovery. By widening his scope and practicing gratitude, he focused on how she directed his attention to seemingly simple things that brought him satisfaction. While on the one hand he had previously taken for granted his wife's "no strings attached" generosity, on the other hand he had fixated on his father's insatiable selfishness. He realized that he had been pouring his efforts down his father's empty drain and taking his wife for granted. With that spark in gratitude for his wife, their relationship shifted. He slowly directed his attention to those people that he had taken for granted. He developed a habit of starting the day by reminding himself of how lucky he was not for his wealth but for his opportunity to use it for social purpose.

Gratitude and Its Benefits

- Highly grateful people compared to their counterparts tend to experience greater life satisfaction and hope (McCullough, Emmons, & Tsang, 2002).
- They also tend to embrace a sense of spirituality, a belief that all things are connected. The combination of mindful focusing on the

area around the heart and intentionally expressing gratitude has been reported to enhance positive emotional states (McGraty & Childre, 2004).

- Gratitude and positive emotional states correlate with relatively more activation of the left prefrontal cortex), anterior cingulate cortex, and pregenual anterior cingulate (Fox, et al.,2015).

Sara was able to tap into Taylor's capacity to be playful that he enjoyed with his wife as a conduit to express generosity and compassion for others. Because he had far more than enough wealth for his family, he slowly learned that the cliché "giving is receiving" did not simply reflect empty platitudes. One of his new hobbies was finding ways to benefit the community as an anonymous benefactor. Just before Christmas he found out where his company's janitors lived and delivered presents to their homes. Not wanting any direct credit for his generosity, he put the presents on their front porches, rang the doorbell, and then ran to hide in the bushes across the street, watching to ensure they answered the door and found the presents. He also stopped going to five-star restaurants and resorts. Instead, he started traveling to third-world countries for a "reality check," ensuring all the while not to behave like an entitled tourist.

Mental Health Benefits of Promoting Gratitude

- Diminished depression
- Diminished anxiety
- Diminished stress
- Increased joy
- Increased enthusiasm
- Increased optimism
- Increased well-being
- Increased life satisfaction

Overall health improves, including enhanced sleep, more exercise, and lower blood pressure.

The prosocial benefits of gratitude include enhancements to romantic relationships, increased social bonds, and the tendency to volunteer more.

By expressing generosity and kindness to others, Taylor promoted positive feedback loops within his relationships, at work, in friendships, and in his marriage. Kindness to others promotes psychological benefits for the giver, including diminished depression, anxiety, and addiction. We can hypothesize that his acts of kindness produced a "helper's high."

Brain Areas Associated With Generosity

- Pleasure centers his brain (Moll, et al., 2006)
- Inferior parietal cortex (Weng, 2013)
- Dorsolateral prefrontal cortex (Weng, 2013)
- Vagus nerve (Steller, et al., , 2015)

These rewards stem from the dopaminergic reward system. A positive emotion "family tree" has been recently proposed, with an array of neurotransmitters stemming from enthusiasm (Shiota et al., 2017). Though emotions cannot be reduced to specific neurotransmitters or hormones, the concept of a spectrum of positive emotions associated with enthusiasm is attractive.

The Dopamine Reward System and Enthusiasm

- Serotonin: pride
- Testosterone: sexual desire
- Oxytocin: nurturant love and contentment
- Cannabinoids: awe and amusement
- Opioids: attachment, gratitude, and pleasure

ACTION-ORIENTED EMPATHY

Cultivating compassion and warm-heartedness for others shifts attention away from narrow interests and suffering (Dalai Lama, 2012). Compassion arises from empathy, as feeling with someone. It is not only wishing to see the person relieved of suffering but also the willingness to do something about it. This distinction describes an essential aspect of psychotherapy because therapists are best at forming an alliance when we express compassion.

Psychodynamic, attachment, and Rogerian theorists have long maintained that when clients feel felt uncritically, they can generate self-compassion. The neurophysiology underlying this feedback loop may involve the anterior cingulate cortex, which is associated with empathy. Expressing compassion generated a positive feedback loop for Taylor, by kindling compassion for himself. Instead of criticizing and judging himself, Taylor took a balanced approach to negative emotions, neither suppressed nor exaggerated, by a willingness to nonjudgmentally accept and acknowledge his humanity.

With self-compassion, the key is the enhancement of self-kindness (Neff, 2011). By doing his best to confront challenges, Taylor behaved in accordance to Reinhold Niebuhr's Serenity Prayer, making a realistic effort each day on what he could do and not wasting energy on trying to do what could not be done, such as pleasing his father. Without this balanced effort, there was always the potential for self-criticism, consistent with a core belief conveyed by his father that he was unworthy. By acknowledging that perfection was not possible, he could make his best effort, and that was "good enough."

Self-criticism is associated with relatively greater activation of the amygdala, right prefrontal cortex, higher cortisol, and increases of adrenaline, as well as with decreased performance and decision making and less resilience following setbacks. Self-compassion is associated with activation of the left prefrontal cortex, insula, decreased cortisol, and increased oxytocin, with clarity of thought and resiliency (Neff, 2011).

Self-compassion is by nature antithetical to narcissism, with benefits to relationships embodied by compassion.

The Benefits of Self-Compassion

- Psychological, such as lower rates of depression, anxiety, recovery from posttraumatic stress disorder, and eating disorders
- Physical, such as alleviating chronic pain, improved lower back pain, and reduced inflammatory response

Self-compassion provided the successful integration of self-referential information to project Taylor into the future with an optimistic bias. The connections between his default-mode and executive networks mobilized his attention and working memory to support a positive stream of thought through the loop between his dorsolateral prefrontal cortex and hippocampus. When he reflected on his past, accessing a library of recent episodes and autobiographical memories to plan for a meaningful future, he moderated the feelings generated by his salience network. These loops operating in his mental networks co-constructed new meaning that embraced gratitude, compassion, and optimism.

Prior to cultivating gratitude, compassion, and forgiveness, Taylor had described himself as fundamentally a pessimist. Yet, he discovered that his old motto, "No good deed goes unpunished," boxed him into an odd form of self-punishment. Whereas pessimism is a retreat, optimism is lean-forward attitude.

The Neuroscience of Optimism

- Increased left prefrontal activation, associated with approach behaviors and positive emotions (Fox, 2013)
- Increased orbitofrontal cortex activation, associated increased resiliency (Dolcos, 2016)
- Enhanced emotion regulation, associated with decreased anxiety and decreased amygdala activation

CONTEMPLATIVE ATTENTION

Ryan had seen a variety of therapists to deal with intermittent bouts of stress and a mix of generalized anxiety and dysthymia. He "never seemed to find the time" to follow through with the lifestyle changes that they recommended. Then a friend invited him to an evening talk on mindfulness, and he was sold. He told his friend, "Who needs all that diet and exercise stuff? If I just give myself a fifteen-minute mindfulness break during my hectic day, and I'm good, right?"

His friend rolled his eyes. "It sounds like if McDonalds had a mindfulness drive-through lane you would think that was enough."

"That's a great idea!" he responded, missing the point.

How does meditation enhance mental health? And how do these practices relate to positive psychology? Despite the recent hype, methods of quiet contemplation, prayer, and meditation have been practiced for at least two thousand of years. Why has there been a consistent tradition? In fact, prior to the emergence of the major theologies, hunter-gatherer societies had shamans, medicine men and women, curanderos, and so on, who may have practiced trance induction, during which they found reverence for their existence. With the advent of the major sociotheological systems twenty-five hundred years ago, contemplative practices were codified into conceptual frames that reflected their host societies (Arden, 1998).

The long history of contemplative meditation in the Christian tradition was practiced with the Desert Fathers in the third century, Saint Augustine and his focus on "the eye of the heart," the Medieval Christian mystics (including Meister Eckhart, Saint Teresa of Avila, and Saint John of the Cross), the monastic orders, and more recently Thomas Merton, who began as a Trappist monk and became an insightful synthesizer of Buddhism and Taoism within the Christian tradition. While the term *meditation* has only recently become popular in Christian circles, the more traditional terms were contemplation and contemplative prayer. In his book *Christian Meditation* (2005), James Finley, a one-time student of Merton's, outlined methods of meditation, including sitting still and upright, eyes closed or lowered, focusing on slow, deep breathing and nonjudgmental compassion. These methods are consistent with those practiced in the East.

Millions of people within the Alcoholics and Narcotics Anonymous programs have been transformed not only through sobriety and social support but also through the belief in a "higher power." At the heart of all major theologies and their respective contemplative practices is the cultivation of meta-awareness, which promotes a soothing sense of security and transcendence.

Over four decades ago, to discover common threads among these methods, I woke before sunrise to meditate in the ashram next door; then, during a year-long trip circling of the globe through Asia, the Middle East, and Europe, I stayed briefly in different religious communes and ashrams, conversing with religious devotees, monks, and priests about their practices. Later studying and then practicing hypnotherapy further reinforced my belief that among all these practices are common factors that promote mental health. The theologically based beliefs, consistent with positive psychology, all seem to cultivate compassion and the belief in interdependence. There are several common methods: hypnosis, contemplative meditation, prayer, and relaxation exercises.

The following factors are common to prayer, meditation, relaxation exercises, and hypnosis. They are calming while reducing the troubling symptoms of psychological disorders and chronic health conditions.

- Deep, gentle, and focused breathing to slow the heartbeat and provide an accessible focal point for attention. The emphasis is on the exhale, emphasizing pressure on the diaphragm, which is associated with stimulating the vagus and the parasympathetic nervous system.
- Shifting attention to the here-and-now, to activate the prefrontal cortex, which calms overreactivity of the amygdala. With a focus on the present moment, the amount of time spent ruminating about the past and worrying about the future shrinks.
- A nonjudgmental attitude helps shift away from rigid expectations to flexible acceptance of whatever occurs.

- Observing body sensations and thoughts allows detachment from pain or general discomfort while simultaneously not denying its existence.
- Labeling anxious and depressing thoughts allows detachment of the thoughts from the emotions.
- A relaxed posture can dissolve body tension, which can be achieved by stretching through traditional yoga or a hybrid yoga.
- A quiet environment provides an opportunity to learn how to relax without distractions.

Contemplative Attention and the Mind's Operating Systems

Meditation, hypnosis, and contemplation can be thought of as attentional methods of orchestrating the balance among the mental networks. Overall, meditation improves attentional flexibility, as the practitioner is less likely to get caught in repetitious ruminative thoughts (Slagter et al., 2007). Like hypnosis, during meditation one disengages attention to fixed narratives generated by excessive default-mode network activity and instead with the executive network promotes nonjudgmental awareness of thoughts, sensations, and emotions as they arise and drift away.

The executive network is critical to the moment-to-moment monitoring of experience during meditation. With the salience network's contribution to subjective awareness of body and emotional feeling states to executive network's attention to decision making and working memory, the variety of mental operating networks training methodologies can be collectively considered as means of contemplative attention. The focus on the breath, a word, or a phrase include broadening the attentional field to observe without judgement. The focused attention is on a calm nonreactive awareness of sensation, emotions, and thoughts as they arise and fade away.

In one study, two types of meditation practices were compared; Kundalini meditation, which involves breath manipulation and an emphasis on somatic awareness, and Tibetan meditation, which emphasizes open monitoring of moment-to-moment experiences. No differences were

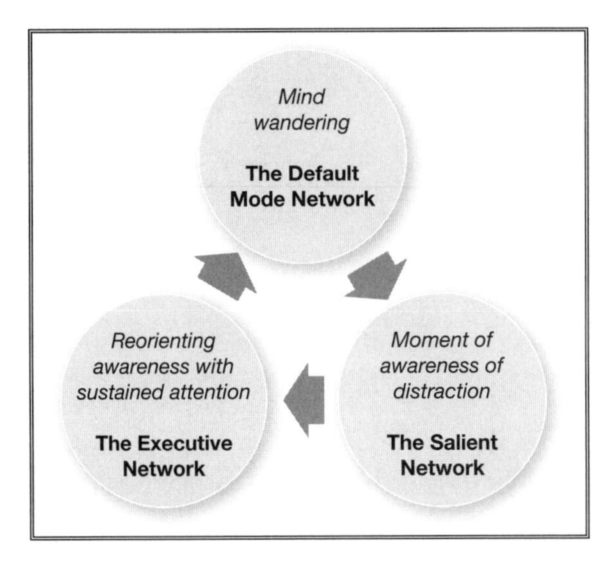

Figure 10.1: The contemplative experiences and the cycle of the activation of the mental operating networks.

found between the meditation groups in heartbeat detection, but they both functioned better than for people in the control group who did not practice meditation (Lutz, Slagter, Dunne, & Davidson, 2008). Both meditation practices provide connectivity between the executive and salience networks through attention to somatic feeling states.

The variety of styles of meditation involve different types of interactions between the mind's operating networks. While the Vipassana tradition is characterized by open monitoring, with the practitioner nonjudgmentally observing moment-to-moment experiences, Zen meditation is characterized by sustained focused attention, such as on a chosen object (Lutz et al., 2008). The impact of meditation training on the activity of the default-mode network during a restful state increases the connectivity with the executive network (Taylor et al., 2013). Self-centered ruminations tend to dissipate with the functional impact of the enhanced connectivity with the default-mode network. With reduced connectivity between the default-mode network and the salience network and greater connectivity with the executive network, one makes less self-referential and emotional judgments based on the misfortunes of the past and instead is more aware of the adaptive potentialities in the future. This amounts to stepping into the future instead of holding on to the past. The greater awareness of the evolving present strengthens affect regulatory capacities while generating detachment from negative feeling states.

Meditation and the Mind's Operating Networks

- The executive network provides a present focus and affect regulation.
- The default-mode network reflects on what we have learned about our place in the world by drawing on autobiographical memories and integrating them in the present moment through the aid of the executive network.
- The salience network allows us to feel ourselves within the world.

A recent study has revealed that creativity increases with the coactivation of the mental operating systems (Beaty, et al, 2018). Normally, the operating systems shift back in forth in their ratio of activity. As we have seen throughout this book, when one network dominates the potential for psychological disorders increases. The balancing of activity increases not only mental health but also creativity.

Contemplative Compassion

Because of the wide-ranging benefits to individuals and societies, the common denominators of all the major theologies, whether on the Judeo-Christian-Islamic spectrum, the Hindu-Jainist-Buddhist spectrum, and the Taoist-Confusionist-Buddhist spectrum, have been the importance of compassion (Arden, 1998). Whether for those in our immediate environment, communities, societies, or internationally, compassion involves leaning toward and expressing kindness. Each of these theologies promotes cultivation of compassion, practiced individually and collectively through contemplative prayer and meditation.

Christian contemplation has been described by Friar Richard Rohr as "the long, loving look at what really is. The essential element in this is time" (2015, pp. 88). The connection between contemplation and compassion is action. Rohr goes on to say, "The effect of contemplation is authentic action, and if contemplation doesn't lead to genuine action, then it remains only self-preoccupation" (p. 121). The concept of compassionate engagement is embraced by all theologies. The emphasis is on action in time, as expressed the saying, "Zen is not like chopping wood. Zen is chopping wood."

My old friend Lisa Leghorn personifies the practice of compassionate activism. In her teens and twenties she was involved with several feminist groups and helped establish one of the first domestic violence shelters in the United States. In her efforts to understand and address the causes of injustice toward women, she moved to Equatorial West Africa for two years to research the social supports for women's economic power in those cultures. Once back in Boston and continuing these lines of inquiry for a book called *Woman's Worth*, her conviction deepened that "economic justice did not go far enough," that broad social transformation requires fundamental changes in consciousness. When I met her forty years ago in Santa Fe she was completing that book while working evenings for Georgia O'Keefe.

Through all these experiences Lisa was becoming increasingly angry and worn down. She explored a number of modalities, searching for ways that would simultaneously sustain herself through activism while making possible a profound change in consciousness she felt was needed. She met H. E. Chagdud Tulku Rinpoche, a Tibetan Buddhist meditation teacher who introduced her to methods for "uprooting negativity" and cultivating the mind's capacity for positive mental habits. First becoming his student, then interpreter, she eventually was ordained by him as Lama Shenpen. She began to teach Bodhisattva peace trainings, where participants from a wide variety of religious backgrounds seeking methods to become more effective in benefiting others are taught to be aware of and transform what is repeated in their minds. Instead of "pushing the pause button" on negative thought patterns, they learn to change those habits to become more fully responsive to the needs of others. Lama Shenpen's search had culminated in the practice of combining meditation, compassion, and activism.

One the methods of meditation she teaches in the Bodhisattva peace training is contemplation, similar to some Christian contemplative methods. First she contemplates compassion, then pauses, "relaxes," and then returns again to contemplate compassion. This paring between the executive network's working memory and attentional capacities and the default-mode network's restful and reflective mode supports homeostatic balance among the operating networks. Contemplation without periodic rest may subtly increase stress. By working toward balance between contemplation and rest, meditation can become increasingly integrated and effortless. Through this rhythm between consuming and digesting, repetition and pacing, the goal is to cultivate compassion that is experiential rather than merely conceptual.

By factoring in self-compassion, contemplative meditation may help balance the interactions between the minds operating networks. One learns to nonjudgmentally observe negative memories and thoughts and to detach negative affect from the thoughts and memories to interrupt the rumination cycle.

Loving kindness meditation, sometimes called metta meditation, involves wishing others well and is associated with the following:

- Increased positive emotions (Frederickson et al., 2008)
- Increase positive feelings of social connection, associated with increased vagal tone (Kok, et al., 2013)
- Reduced migraines (Tonelli, & Wachholtz, 2014))
- Improved lower-back pain (Carson, 2005)
- Decreased symptoms of posttraumatic stress syndrome in veterans (Kearney, et al., 2013)

Meditation-Augmented Therapy

To what degree do meditative practices contribute to the mental health? With all the recent attention in either applying or combining meditation to psychotherapy, it is important to consider the factors that increase positive outcomes. Mindfulness-based stress reduction (MBSR) has been studied extensively for its utility in helping people with general and mental health problems. MBSR has been shown to reduce amygdala activity in people with social anxiety disorder (Goldin & Gross, 2010). Accordingly, it is associated with decreases in negative emotions and increases in attention-related neural networks.

Neural Correlates of Mediation

- Stress reduction through MBSR training has been correlated with incremental increases in antibody response (Davidson et al., 2009).
- MBSR training over time has been associated with reductions in gray matter densities in the right basolateral amygdala (Holzel et al., 2010).
- Long-term meditators show increased thickness of the medial prefrontal cortex and also enlargement of the right insula (Lazar et al., 2005). The medial prefrontal cortex has been associated with self-observation connected with mindfulness meditation (Cahn & Polich, 2006).
- Long-term meditation has been shown to increase thickness of the prefrontal cortex, which augments affect regulation, observance, and attention. The increased thickness of the right insula corresponds to heightened awareness of the mediator's own body.
- Meditation has been associated with significant effects in neuroplasticity of the left prefrontal cortex and decreased amygdala response to stress (Davidson, et al., 2003).
- Memory and learning centers associated with the left hippocampus can change within eight weeks of practice (Holzel et al., 2011).
- With elevated activity in the dorsolateral prefrontal cortex and rostral anterior cingulate cortex with a corresponding decrease of

amygdala activity and grey matter density, there are offsets to cortical thinning (Lazar, 2005).
- Mindfulness training is associated with greater left prefrontal cortex activation, stress reduction benefits, and increases in antibodies to influenza vaccine (Davidson et al., 2003).

One of the main common denominators among MBSR approaches involves moderating the stress response systems while activating the parasympathetic nervous system. Through activities such as yoga, meditation, and tai chi, improvements have been reported in mood and boosts the immune system have been measured by decreased production of interferon-gamma and increased production of interleukin-4, an anti-inflammatory cytokine, by stimulated T-cells (Carlson et al., 2003).

Since the late 1960s researchers and practitioners had been using biofeedback instruments to demonstrate how people attain deeper states of relaxation. More recently, there has been an intensified focus with neurofeedback instruments. Long-term meditation practitioners have been observed to produce marked increases in the brain's electrical signals as measured in the fast-frequency oscillation referred to as gamma waves, particularly in the prefrontal cortex (Davidson et al., 2003). In contrast to the assumption that meditation is analogous to simple relaxation, the increase in gamma waves represents an increase in synchrony between the prefrontal cortex and other brain regions not associated with relaxation. Gamma and beta brain oscillations are associated with attention, perception, and learning. Synchronized activity that selects information for further processing has a much stronger impact on cells than temporarily uncoordinated activity associated with relaxation. Attention synchronization increases learning. A person may "learn" to be phobic and need to learn not to be phobic. Through relearning this new synchronized state rewires the brain. Meditation offers one of the many ways that can facilitate a new synchronized state that can augment psychotherapy to ameliorate anxiety and depression.

Mind-body programs have become a staple for many medical centers. Twenty years ago at Kaiser Permanente my mind-body class was one of many. Each of our twenty-four medical centers in Northern California had mindfulness classes. Because mind-body medicine has become mainstream, we are able to gain from large population studies throughout the world. For example, the effect on anxiety and depression was examined in a meta-analysis of 39 studies totaling 1,140 participants receiving mindfulness-based therapy for a range of conditions such as cancer, generalized anxiety disorder, and depression. The overall results indicated that mindfulness-based therapy was moderately effective for improving anxiety and mood symptoms (Hofmann, et al., 2010). Positive effects on anxiety, depression, and pain were found by a systematic study of 47 randomized controlled trials, with a total of 3,515 participants, but the effects were not superior to exercise, yoga, muscle relaxation, cognitive behavioral therapy, or medication (Goyal et al., 2014).

The effects of a variety of methods including tai chi, quigong, meditation, and yoga on the immune system were examined in comprehensive meta-analysis of randomized controlled trials. The immune-system-related inflammatory outcomes indicated a moderate reduction of C-reactive protein, a small but nonsignificant reduction of interleukin-6, and no effect on tumor necrosis factor-α, immune factor CD4, or natural killer cell counts (Morgan, Irwin, Chung, & Wang, 2014). There were stronger effects from tai chi, quigong, and exercise compared to meditation, perhaps from the benefit of movement. In summary, mind-body therapies can influence some aspects of the immune system, including some inflammatory measures, but again, they are incomplete by themselves.

A "third wave" of cognitive behavioral therapy has included a mindfulness component. The most prominent of these include mindfulness-based cognitive therapy, dialectical behavior therapy, and acceptance commitment therapy. These approaches teach people how to relate differently to their thoughts, feelings, and bodily sensations. Borrowing from Buddhism but not dissimilar to Christian contemplative methods, they promote accepting thoughts and feelings without judgment. Rather than trying to push the thoughts and feelings out of consciousness, a client may be able to indirectly reduce automatic and maladaptive reactions to thoughts, feelings, and events. The differences between these metacognitive models and traditional cognitive behavioral therapy approaches typify the following exchange: When depressed individuals say, "I am a bad and defective person," cognitive behavioral therapists ask them to challenge the validity of that statement and develop alternative thoughts. In contrast, metacognitive-oriented therapists teach them to say: "I'm just having the thought that I'm a bad and defective person," or "It is just a thought. It does not mean that I am in fact a bad person." This slight change is meant to detach the thought from the emotion and reduce the destructive power of the negative thought. Labeling emotions to reduce anxiety and depression tends to inhibit tame the amygdala through the prefrontal cortex's emotional regulation pathways.

Mindfulness-Augmented Therapy

The adoption of mindfulness-based approaches within these therapeutic modalities has been applied to specific disorders:

- Anxiety disorders, with acceptance commitment therapy (Hays & Strosahl, 2016)
- Borderline personality disorder, within the dialectical behavior therapy (Linehan, 1993)
- Obsessive-compulsive disorder, with mindfulness added to cognitive behavioral therapy (Baxter et al., 1992),
- Depression, with mindfulness-based cognitive behavioral therapy (Teasdale et al., 2000; Ma & Teasdale, 2004)
- General medical problems, such as chronic pain (Kabat-Zinn, 1982)

The practices of contemplative meditation and mindfulness attempt to look for novelty, subtle beauty, and satisfaction in each moment by activating attentional networks. This open focus strengthens the indirect dopamine pathways to inhibit impulsiveness while enlarging the scope of potential pleasurable experiences. With less impulsive habit-driven behaviors, one is better able to learn to appreciate the rich complexity of life. Instead of being disturbed ambiguity, one is fascinated by it. Much of the movement toward greater "self"-organization depends on the ability to sustain attention and appreciation for the subtleties of present moment.

THE "SELF"-ORGANIZING MIND

This book has explored the many feedback loops that contribute to mental health or ill health. The quest to find "the" factor that "causes" poor mental health or the "correct" therapy has led to dead ends. The belief in the pure essence of things, including our "core identity" as individuals, has faded as we have come to understand that DNA is not destiny (Heine, 2017). Research in epigenetics and psychoneuroimmunology has shown that genes can turn on or off and that disease processes and mental health are significantly related to our lifestyle and experience. Our brain and stress response systems adapt to our experiences. And our minds are influenced by all these interdependent systems, not a static self.

We use our mental operating networks to "self"-organize. As complex adaptive systems, we self-organize in time, neither staying completely the same nor changing completely. We can evolve and become progressively more adaptive and healthy or devolve and become ill. Time, and ourselves within it, does not stand still. Healthy minds adapt to changing environments and create situations that promote health.

Feelings play a role in the awareness of time, self, and the global emotional moment of subjective time. This subjective awareness feedback loop "self"-organizes increasing scales in complexity of a sense of self. Rather than having a static sense of self, in therapy clients learn to identify perceptual feelings of themselves across time that contribute to the subjective awareness of a sense of "me" in the ongoing present moment.

Self-organization maximizes the flexible zone between stability and change. While too much stability causes rigidity and leads to depression, too much change is chaotic and leads to anxiety. Healthy 'self"-organization is both self-referential for stability and adaptive so that we become something greater than we had been. The mind is a self-organizing emergent process that is both influenced by and influencing its subordinate factors (Arden, 1996; Siegel, 2017). Given that the mind is a self-organizing process, its self-recursive shaping of information contributes to "self"-organization of our evolving individuality, neither static and unchanging nor always in flux.

Complex adaptive systems, such as ourselves, are composed of multiple feedback loops. The more complex the feedback loops, the more complex the adaptive system's ability to "look ahead" in time, examining the effects of various action sequences without executing actions.

As complex adaptive systems, we thrive with openness and flexibility, maintaining continuity though self-recursive self-organization. Healthy minds are not unconstrained by chaos or rigidity that impedes growth. Unconstrained chaos in psychological terms represents anxiety, so that the person who feels out of control, activating stress systems when there is no real danger. A depressed person's rigidity, being stuck in the past, ruminating over negative events in his life, holds onto the same negativistic story line and cultivates a depressed mood. When healthy, we are flexible and adaptable and can deal with near chaotic situations by maintaining internal stability and continuity.

Because we are complex adaptive systems, we are part of the environment (space) while adapting to our era (time). Though as Wittgenstein noted there are limitations of language, we may clumsily say that the brain is a noun (a substance) and that the mind is a verb (a process). Our minds, composed of mental operating networks and memory systems, adapt to the meanings we construct and so modify our brains and bodies.

As complex adaptive systems, the interactions among our subsystems (genes, immune systems, bodies, etc.) give rise to the emergent mental networks that represent the superordinate capacity to think, feel, sense, and behave in extraordinarily complex ways. We develop relationships with other minds with the same capacities, leading to more complex relationships within families, towns, cities, and societies. All those mutually interdependent factors self-organize through openness and nonlinearity.

The mind is an emergent "self"-organizing process that cannot be reduced to its constituent parts, its operating networks. It is profoundly influenced by all elements of the self—genes, immune and neural systems, family dynamics, and culture—and it influences each of these systems at various levels. From this point of view, the mind is a meaning-making process constructed from the nonlinear interactions among all of its contributing subsystems. Psychotherapy can help orchestrate these changes to promote a healthy mind-body by reintegrating the subsystems that had become dysregulated. Therapy must take into account that as complex systems we need to facilitate the flexible interaction among the mind's operating networks. Within the relevant cultural framing and socioeconomic position, psychotherapy in the twenty-first century must address all these feedback loops.

In the one hundred years since the theologian and paleontologist Pierre Theilard de Chardin wrote *The Phenomenon of Man*, we have come to the point in evolution where we can understand the emergence of minds. In so doing, we are in a position to offer help to people whose minds work against them, in a far more integrated way than ever before. And so the practice of psychotherapy connects the domains that previously had been compartmentalized. The feedback loops within the mind-brain-gene continuum will continue to be articulated in the years to come, so that we can truly offer an integrated psychotherapy in the twenty-first century.

References

Adams, W. K., Sussman, J. L., Kaur, S., D'souza, A. M., Kieffer, T. J., & Winstanley, C. A. (2015). Long-term, calorie-restricted intake of a high-fat diet in rats reduces impulse control and ventral striatal D2 receptor signaling: Two markers of addiction vulnerability. *European Journal of Neuroscience, 42*(12), 3096–3104.

Adlard, P. A., Perreau, V. M., & Cotman, C. W. (2005). The exercise-induced exercise-induced expression of BDNF within the hippocampus varies across life-span. *Neurobiology of Aging, 26,* 511–520.

Allen, J. G. (2001). *Traumatic relationships and serious mental disorders.* New York, NY: John Wiley and Sons.

Allman, J., Tetreault, N. A., Hakeem, A. Y., & Park, S. (2011). The von Economo neurons in apes and humans. *American Journal of Human Biology, 23,* 5–21.

Allman, J., Watson, K. K., Tetreault, N. A., & Hakeem, A. Y. (2005). Intuition and autism: A possible role for von Economo neurons. *Trends in Cognitive Neuroscience,* 9 367–373.

Amara, R., Bodenhorn, K., Cain, M., Carlson, R., Chambers, J., Cypress, J., Cypress, . . . , Yu, K. (2003). *Health and healthcare 2010: The forecast and challenge.* San Francisco, CA: Jossey-Bass.

Anstey, K. J., Cherbuin, N., Budge, M., & Young, J. (2011). Body mass index in midlife as a risk factor for dementia: A meta-analysis of prospective studies. *Obesity Reviews, 12,* 426–437.

Arden, J. B. (1996). *Consciousness, dreams, and self: A transdisciplinary perspective.* Madison, CT: International Universities Press.

Arden, J. B. (1998). *Science, theology, and consciousness.* Westport, CT: Praeger.

Arden, J. B., (2009a) *Brain-Based Therapy-Child* John Wiley and Sons.

Arden, J. B., (2014) *The Brain Bible.* New York, NY: McGraw Hill.

Arden, J. B. (2015). *Brain2Brain*. New York, NY: Wiley.

Ashman, S. B., Dawson, G., Panagiotides, H., Yamada. E, Wilkinson, C. W. (2002) Stress hormone levels of children of depressed mothers. *Developmental Psychopathology, 14*(2), 333–49.

Azar, B. (2011). Oxytocin's other side. *APA Monitor on Psychology, 42,* 40.

Bandura, A. (1997) Self-efficacy: the exercise of control. W. H. Freeman and Company.

Baran, S. E., Campbell, A. M., Kleen, J. K., Foltz, C. H., Wright, R. L., Diamond, D. M., & Conrad, C. D. (2005). Combination of high fat diet and chronic stress retracts hippocampal dendrites. *Neuroreport, 16*(1), 39–43.

Barnett, L. (2017). *How emotions are made.* New York, NY: Houghton Mifflin Harcourt.

Barr, C. S., Schwandt, M. L., Lindell, S. G., Higley, J. D., Maestripieri, D., Goldman, D., . . . Heilig, M. (2008). Variation at the mu-opioid receptor gene (OPRM) influences attachment behavior in infants. *Proceedings of the National Academy of Sciences of the USA, 105,* 5277–5281.

Bassuk, S. S., Glass, T. A., Berkman, L. F. (1999). Social engagement and incident cognitive decline in community – dwelling elderly persons. *Annuals of Internal Medicine, 131,* 165–173.

Baxter, L. R., Jr, Schwartz, J. M., Bergman, K. S., Szuba, M. P., Guze, B. H., Mazziotta, J. C., et al. (1992) Caudate glucose metabolic rate changes with both drug and behavior therapy for obsessive–compulsive disorder. *Archives of General Psychiatry,* 49, 681–9.

Baylin, J., & Hughes, D. (2016). *The neurobiology of attachment-focused therapy: Enhancing connection and trust in the treatment of children and adolescents.* New York, NY: Norton.

Beaty, R. E., Kenett, Y. N., Christensen, A. P., Rosenberg, M. D. Benedek, M., Chen, Q., Fink, A., Qiu, J., Kwapil, T. R., Kane, M. J., & Silvia (2018) Robust prediction of individual creative ability from brain functional connectivity. *Proceedings of the National Academy of Sciences, 115* (5), 1087-1092; DOI:10.1073/pnas.1713532115

Beebe, B., Jaffe, J., Markese, S., Buck, K., Chen, H., Cohen, P., . . . Feldstein, S. (2010). The origins of 12-month attachment: A microanalysis of 4-month mother-infant interaction. *Attachment and Human Development, 12*(1–2), 3–141. http://doi.org/10.1080/14616730903338985

Beebe, B., & Lachmann, F. (2002). *Infant research and adult treatment: Co-constructing interactions.* Hillsdale, NJ: Analytic Press.

Benedetti, F. (2008). *Placebo effects: Understanding the mechanisms in health and disease.* New York, NY: Oxford University Press.

Benros, M. E., Waltoft, B. L., Nordentoft, M., Ostergaard, S. D., Eaton, W. W., Krogh J., & Mortensen P. B. (2013). Autoimmune disease and severe infections as risk factors for mood disorders: A nationwide study, *JAMA Psychiatry, 70*(8), 812–20. http://doi.org/10.1001/jamapsychiatry.2013.111

Bercik, P., Denou, E., Collins, J., Jackson, W., Lu, J., Jury, J., (. . .) Collins, S. M. (2011). The intestinal microbiota affect central levels of brain-derived neurotrophic factor and behavior in mice. *Gastroenterology, 141*, 599–609.

Berridge, K. C. (2009). Wanting and liking: Observations from the neuroscience and psychology laboratory, *Inquiry, 52*(4), 378-398, DOI: 10.1080/00201740903087359.

Berridge, K. C., & Kringelbach, M. L. (2008). Affective neuroscience of pleasure: Reward in humans and animals. *Psychopharmacology, 199*, 457–480. http://doi.org/10.1007/s00213-008-1099-6

Bierhaus, A., Humpert, P., & Nawroth, P. (2006). Linking stress to inflammation. *Anesthesiology Clinics, 24*, 325–340.

Blackburn, E. & Epel, E. (2017) *The telomere effect: A revolutionary approach to living younger, healthier, longer.* New York: Grand Central Publishing

Blandino, P., Jr., Barnum, C. J., & Deak, T. (2006). The involvement of norepinephrine and microglia in hypothalamic and spleen IL-6 beta responses to stress. *Journal of Neuroimmunology, 173*(1-2), 87–95. https://doi.org/10.1016/j.jneuroim.2005.11.021

Bluhm, R., Williamson, P., Lanius, R., Theberge, J., Densmore, M., Bartha, R., . . . Osuch, E. (2009). Resting state default-mode network connectivity in early depression using a seed region-of-interest analysis: Decreased connectivity with caudate nucleus. *Psychiatry and Clinical Neuroscience, 63*, 754–761.

Boccardi, V., Esposito, A., Rizzo, M. R., Marfella, R., Barbieri, M., & Paolisso, G. (2013). Mediterranean diet, telomere maintenance and health status among elderly. *PLOS One*, 8(4), e62781. http://doi.org/10.1371/journal.pone.0062781

Booth, F. W., & Neufer, P. D. (2005). Exercise control gene expression: The activity level of skeletal muscle modulates a range of genes that produce dramatic molecular changes—and keeps us healthy. *American Scientist, 93*, 28-35.

Borysenko, J. (2014). *The plants plus diet.* New York, NY: Hay House.

Bouvard, V., Loomis, D., Guyton, K. Z., Grosse, Y., Ghisassi, F. E., Benbrahim-Tallaa, L., Guha, N., Mattock, Heidi, & Staif, K. (2015). Carcinogenicity of consumption of red and processed meat. *The Lancet Oncology, 16*, 1599–1600.

Braver, T. S., Barch, D. M., Gray, J. R., Molfese, D. L., & Synder, A. (2001). Anterior cingulate cortex and response conflict: Effects of frequency, inhibition and errors. *Cerebral Cortex, 11*, 825–836.

Briere, J., & Scott, C. (2015). *Principles of trauma therapy: A guide to symptoms, evaluation, and treatment.* Thousands Oaks, CA: Sage Publications.

Brown, A. S., van Os, Driessens, C, Hoek, H.W, & Susser, E.S. (2000). Further evidence of relationship between prenatal famine and major affective disorder. *American Journal of Psychiatry, 157*(2), 190–195.

Brown, G. W. (1998). Genetic and population perspectives on life events and depression. *Social Psychiatry and Psychiatric Epidemiology, 33*(8), 363–372.

Brunn, J. M., Lihn, A. S., Verdich, C., Pederson, S. B., Toubro, S., Astrup, A., & Richelsen B. (2003). Regulation adiponectin by adipose tissue-derived cytokines: In vivo and in vitro investigations in humans. *American Journal of Physiology, Endocrinology, and Metabolism, 285,* E527–E533.

Buckner, R. L., Andrews-Hanna, J. R., & Schacter, D. L. (2008). The brain's default mode network: Anatomy, function, and relevance to disease. *Annals of the New York Academy of Science, 11,* 49–57.

Burgdorf, J., Kroes, R. A., Beinfeld, M. C., Panksepp, J., & Moskal, J. R. (2010). Uncovering the molecular basis of positive affect using rough-and-tumble play in rats: A role for insulin-like growth factor 1. *Neuroscience, 163,* 769–777.

Burke, B. L., Arkowitz, H., & Menchola, M. (2003). The efficacy of motivational interviewing: A meta-analysis of controlled clinical trials. *Journal of Consulting and Clinical Psychology, 71*(5), 843-861.

Butler, R. N., Forette, F., & Greengross, B. S. (2004). Maintaining cognitive health in an ageing society. *Journal of the Royal Society of Health, 124*(3), 119–121.

Cahill S. P., Carrigan M. H., Frueh B. C. (1999). Does EMDR work? And if so, why?: A critical review of controlled outcome and dismantling research. *Journal of Anxiety Disorders, 13,* 5–33. 10.1016/S0887-6185(98)00039-5

Cai, N., Chang S., Li Y., Li Q., Hu J., Liang J., Flint J. (2015). Molecular signatures of major depression. *Current Biology, 25*(9), 1146–1156. http://doi.org/10.1016/j.cub.2015.03.008

Caldji, C., Diorio, J., & Meaney, M. J. (2003). Variations in maternal care alter GABA-A receptor subunit expression in brain regions associated with fear. *Neuropsychopharmacology, 28,* 150–159.

Callaghan, B. L., Sullivan, R. M., Howell, B., & Tottenham, N. (2014). The international society for the Developmental Psychoneurobiology Sackler Symposium: Early adversity and the maturation of emotion circuits—a cross-species analysis. *Developmental Psychobiology, 56*(8), 1635–1650. http://doi.org/10.1002/dev.21260

Cannon, W. (1989). *The wisdom of the body*. Birmingham, AL: Classics of Medicine Library. (Original work published 1939)

Carlson, L.E., Speca, M., Patel, K.D., & Goodey, E. (2003) Mindfulness-based stress reduction in relation to quality of life, mood, symptoms of stress, and immune parameters in breast and prostate cancer outpatients. Psychosom Med, 65(4), 571–581.

Carpenter, Tyrka A. R., McDougle C. J., Malison R. T., Owens M. J., Nemeroff C. B., & Price, L. H. (2004). Cerebrospinal fluid corticotropin-releasing factor and perceived early-life stress in depressed patients and healthy control subjects. *Neuropsychopharmacology, 29*, 777–784.

Carr, J. (1996) Neuroendocrine and behavioral interaction in exposure treatment of phobic avoidance, *Clinical Psychology Review, 16*(1), 1–15.

Carro, E., Trejo, J. L., Busiguina, S., & Torres-Aleman, I. (2001). Circulating insulin-like growth factor 1 mediates the protective effects of physical exercise against brain insults of different etiology and anatomy. *Journal of Neuroscience, 21*, 5678–5684.

Caspi, A., McClay, J., Moffitt, T. E., Mill, J., Martin, J., Craig, I. W., . . . Poulton, R. (2002). Role of genotype in the cycle of violence in maltreated children. *Science, 297*, 851–854.

Caspi, A., Moffitt, T. E., Cannon, M., McClay, J., Murray, R., Harrington, H., . . . Craig, I. W. (2005). Moderation of the effect of adolescent-onset cannabis use on adult psychosis by a functional polymorphism in the COMT gene: Longitudinal evidence of gene x environment interaction, *Biological Psychiatry, 57*, 1117–1127.

Cassidy, J., Jones, J. D., & Shaver, P. R. (2013). Contributions of attachment theory and research: A framework for future research, translation, and policy. *Development and Psychopathology, 25*(402), 1415–1434.

Champagne, F. A., & Curley, J. P. (2009). Epigenetic mechanisms mediating the long-term effects of maternal care on development. *Neuroscience and Biobehavioral Reviews, 33*(4), 593–600.

Champagne, F. A., Weaver, I. C., Diorio J, Dymov S, Szyf M, & Meaney MJ.. (2006). Maternal care associated with methylation of the estrogen receptor-alpha16 promoter and estrogen receptor-alpha expression in the medial preoptic area of female offspring. *Endocrinology, 147*(6), 2909–2915.

Chen, E., Miller, G. E., Walker, H. A., Arevalo, J. M., Sung, C. Y., & Cole, S. W. (2007). Genome-wide transcription profiling linked to social class in asthma. *Thorax*, 64(1), 38–43. doi: 10.1136/thx.2007.095091.

Chiron, C., Jambaque, I., Nabbout, R., Lounes, R., Syrota, A., & Dulac, O.

(1997). The right hemisphere is dominant in human infants. *Brain, 120,* 1057–1065. http://doi.org/10.1093/brain/120.6.1057

Cirelli C., & Tononi G. (2008) Is sleep essential? PLoS Biology, 6(8), e216.

Coan, J. A., Schaefer, H. S., & Davidson, R. J. (2006). Lending a hand: Social regulation of the neural response to threat. *Psychological Science, 17,* 1032–1039.

Cohen S., Doyle W.J., Skoner D.P., Rabin B.S., & Gwaltney J.M. Jr. (1997) Social ties and susceptibility to the common cold. *JAMA, 277,* 1940 –1944.

Cohen S. (2004) Social relationships and health. *American Psychologist, 59,* 676–684.

Cole, S. W., Hawkley, L. C., Arevalo, J. M., Sung, C. Y., Rose, R. M., & Cacioppo, J. T. (2007). Social regulation of gene expression in human leukocytes. *Genome Biology, 8,* R189.

Comings, D. E., & Blum, K. (2000). Reward deficiency syndrome: Genetic aspects of behavioral disorders. *Progress in Brain Research, 126,* 325–341.

Cordain, L., Gotshall, R. W., & Eaton, S. B., III (1998). Physical activity, energy expenditure, and fitness: An evolutionary perspective. *International Journal of Sports Medicine, 19,* 328–335.

Cozolino, L. (2014). *The neuroscience of human relationships: Attachment and the developing social brain* (2nd ed.). New York, NY: Norton.

Cozolino, L. (2017). *The Neuroscience of psychotherapy: Healing the social brain* (3rd ed.). New York, NY: Norton.

Craig, A. D. (2015). *How do you feel? An interoceptive moment with your neurobiological self.* Princeton, NJ: Princeton University Press.

Curley, J. P. (2011). The mu-opioid receptor and the evolution of mother-infant attachment: Theoretical comment on Higham et al. *Behavioral Neuroscience, 125,* 273–278. http://doi.org/10.1037/a0022939

Czeh, B., Muller-Keuker, J. I., Rygula, R., Abumaria, N., Hiemke, C., Domenici, E., & Fuchs, E. (2007). Chronic social stress inhibits cell proliferation in the adult medial prefrontal cortex. *Neuropsychopharmacology, 32*(7), 1490–1503.

Dalai Lama (2012). *Beyond religion.* Wilmington, MA: Mariner Books.

Damasio, A. (1994). *Descartes error: Emotion, reason, and the human brain.* New York, NY: Picador.

Danese, A., Pariante, C. M., Caspi, A., Taylor, A., & Poulton, R. (2007). Childhood maltreatment predicts adult inflammation in a life-course study. *Proceedings of the National Academy of Sciences of the USA, 104,* 1319–1324.

Dantzer, R., Kelley K. W. (2007) Twenty years of research on cytokine-induced sickness behavior. *Brain Behavior Immunity, 21*(2), 153–160.

Davidson R.J., Kabat-Zinn J., Schumacher J., Rosenkranz M., Muller D., Santorelli S.F., Urbanowski F., Harrington A., Bonus K., Sheridan J.F. (2003) Alterations in brain and immune function produced by mindfulness meditation. *Psychosomantic Medicine*, 65(2), 564–570. doi: 10.1097/01 .PSY.0000077505.67574.E3.

Dawes, C. T., Loewen. P. J. Shreiber, D., et al. (2012). Neural basis of egalitarian behavior. *Proceedings of the National Academy of Sciences of the USA*, *109*(17), 6479–6483.

De Havas, J. A., Parimal, S., Soon C. S., & Chee, M. W. (2012). Sleep deprivation reduces default mode network connectivity and anti-correlation during rest and task performance. *NeuroImage*, *59*(2), 1745–1751. http://doi.org/10.1016/j.neuroimage.2011.08.026

Depue, R. A., & Morrone-Strupinsky, J. V. (2005). A neurobehavioral model of affiliative bonding: implications for conceptualizing a human trait of affiliation. *Behavioral and Brain Sciences*, *28*(3), 313–395.

Ding J, Gip P, Franken P, Lomas L, & O'Hara B. (2004) A proteomic analysis in brain following sleep deprivation suggests a generalized decrease in abundance for many proteins. *Sleep*; *27*, A391.

Domes, G., Heinrichs, M., Berger, C., & Herpertz, S. C. (2007). Oxytocin improves "mind-reading" in humans. *Biological Psychiatry*, *61*, 731–733.

Dong, Y., Green, T., Saal, D., Marie, H., Neve, R., Nestler, E. J., & Malenka, R. C. (2006). CREB modulates excitability of nucleus accumbens neurons. *Health Services Research*, *41*(6), 2255–2266.

Engert, V., Joober, R., et al. (2009). Behavioral response to methylphenidate challenge: Influence of early life parental care. *Developmental Psychobiology*, *51*(5), 408–416.

Epel, E. S., Blackburn, E. H., Lin, J., Dhabhar, F. S., Adler, N. E., Morrow, J. D., & Crawthon, R. M. (2004). Accelerated telomere shortening in response to life stress. *Proceedings of the National Academy of Sciences of the USA*, *101*, 17312–17315.

Etkin, A., Egner, T., Peraza, D., Kandel, E., & Hirsch, J. (2006). Resolving emotional conflict: A role for the rostral anterior cingulate cortex in modulating activity in the amygdala. *Neuron*, *51*, 871–882

Evans, J. A., & Davidson, A. J. (2013). Health consequences of circadian disruption in humans and animal models. *Progress in Molecular Biology and Translational Science*, *119*, 283–323.

Fabel, K., Fabel, K., Tam, B., Kaufer, D., Baiker, A., Simmons, N., et al. (2003). VEGF is necessary for exercise-induced adult hippocampus neurogenesis. *European Journal of Neuroscience*, *18*(10), 2803-2812.

Fang, Z., Spaeth, A. M., Ma, N., Zhu, S., Hu, S., Goel, N., . . . Rao, H. (2015).

Altered salience network connectivity predicts macronutrient intake after sleep deprivation *Scientific Reports, 5,* article 8215.

Feart, C., Samieri, C., Rondeau, V., Amieva, H., Portet, F., Dartigues, J. F., . . . Barberger-Gateau, P. (2009). Adherence to a Mediterranean diet, cognitive decline, and risk of dementia. *JAMA, 302*(6), 638–648.

Felitti, V., Anda, R. F., Nordenberg, D., Williamson, D. F., Spitz, A. M., Edwards, V., et al. (2001). Relationship of childhood abuse and household dysfunction to many of the leading causes of death in adults. In K. Franey, R. Geffner, & R. Falconer (Eds.), *The cost of child maltreatment: Who pays? We all do* (pp. 53–69). San Diego, CA: Family Violence and Sexual Assault Institute.

Felitti, V. (2009) Adverse childhood experiences and adult health. *Academy of Pediatrics,* 9, 131-132.

Ferber S.G., Feldman R., & Makhoul, I.R. (2008) The development of maternal touch across the first year of life. *Early Human Development, 84,* 363–370

Ferrie, J. E., Shipley, M. J., Akbaraly, T. N., Marmot, M. G., Kivimäki, M., & Singh-Manoux, A. (2011). Change in sleep duration and cognitive function: Findings from the Whitehall II Study. *Sleep, 34*(5), 565–573.

Field, C. B., Johnston, K., Gati, J. S., Menon, R. S., & Everling, S. (2008). Connectivity of the primate superior colliculus mapped by concurrent microstimulation and event-related FMRI. *PLOS One, 3*(12), e3928.

Finley, J. (2005). *Christian meditation.* New York, NY: Harper Collins.

Fiori, L. M., & Turecki, G. (2012). Broadening, our horizons: Gene expression profiling to help better understand the neurobiology of suicide and depression. *Neurobiology Disorders, 45,* 14–22.

Fitzgerald, P. B., Laird, A. R., Maller, J., & Daskalakis, Z. J. (2008). A meta-analytic study of changes in brain activation in depression. *Human Brain Mapping, 29,* 683–695.

Flanders, J. L., Leo, V., Paquette, D., Robert O. Pihl, R. O.,& Jean R. Séguin, J. R. (2009). Rough-and-tumble play and the regulation of aggression: An observational study of father–child play dyads. *Aggressive Behavior, 35*(4), 285–295.

Foley, D. L., Eaves, L. J., Wormley, B., Silberg, J. L., Maes, H. H., Kuhn, J., & Riley, B. (2004). Childhood adversity, monoamine oxidase A genotype, and risk for conduct disorder. *Archives of General Psychiatry, 61,* 763–744.

Fox, G. R., Kaplan, J., Damasio, H., & Damasio, A. (2015). Neural correlates of gratitude. *Frontiers in Psychology,* 6, 1491. http://doi.org/10.3389/fpsyg.2015.01491

Francis, D. D., & Meaney, M. J. (1999). Maternal care and the development of stress responses. *Current Opinion in Neurobiology,* 9(1), 128–134.

Francis, D. D., & Kuhar, M. J. (2008). Frequency of maternal licking and grooming correlates negatively with vulnerability to cocaine and alcohol use in rats. *Pharmacology Biochemistry, & Behavior, 90,* 497–500. doi:10.1016/j.pbb.2008.04.012

Frank, M. G., Issa, N. P., & Stryker, M. P. 2001. Sleep enhances plasticity in the developing visual cortex. *Neuron, 30,* 275–287.

Fredrickson, B. L., Cohn, M. A., Coffey, K. A., Pek, J., & Finkel, S. M. (2008). Open hearts build lives: Positive emotions, induced through loving-kindness meditation, build consequential personal resources. *Journal of Personality and Social Psychology,* 95(5), 1045–1062. http://doi.org/10.1037/a0013262

Frodl, T., Reinhold, E., Koutsouleris, N., et al. (2010). Childhood stress, serotonin transporter gene and brain structures in major depression. *Neuropsychopharmocology, 35,* 1383–1390.

Fuster, J. (2008). *The prefrontal cortex* (4th ed.). Cambridge, MA: Academic Press.

Gilbert, D. (2007). *Stumbling on happiness.* New York, NY: Vintage.

Gimpli, G., & Fahrenholz, F. (2001). The oxytocin receptor system: Structure, function, and regulation. *Physiological Reviews, 81,* 629–683.

Glaus, J., et al. (2014). Association between mood, anxiety, or substance use disorders and inflammatory markers after adjustment for multiple covariates in a population-based study. *Journal of Psychiatric Research, 58,* 36–45. http://doi.org/10.1016/j.psychires.2014.07.012

Goebel, M. V., Mills, P. J., Irwin, M. R., & Ziegler, M. G. (2000). Interleukin-6 and tumor necrosis factor-alpha production after acute psychological stress, exercise, and infused isoproterenol: Differential effects and pathways. *Psychosomatic Research, 62,* 591–598.

Goldin, P. R., & Gross, J. J. (2010). Effects of mindfulness-based stress reduction (MBSR) on emotion regulation in social anxiety disorder. *Emotion, 10*(1), 83–91. http://doi.org/10.1037/a0018441

Goldin, P., Ziv, M., Jazaieri, H., Werner, K., Kraemer, H., Heimberg, R. G., & Gross, J. J. (2012). Cognitive reappraisal self-efficacy mediates the effects of individual cognitive-behavioral therapy for social anxiety disorder. *Journal of Consulting and Clinical Psychology,* 80(6), 1034–1040. http://doi.org/10.1037/a0028555

Goldstein, N., Peretz, I., Johnsen, E., & Adolphs, R. (2007). The orbitofrontal cortex in drug addiction. In D. H. Zald & Rauch, S. L. (Eds.), *The orbital frontal cortex* (pp. 481–522). New York, NY: Oxford University Press.

Goodwin, R. D., Wickramaratne, P., Nomura, Y., & Weisman M. M. (2007). Familial depression and respiratory illness in children. *Archives of Pediatrics and Adolescent Medicine, 161*(5), 487–494.

Gordon, N. S., Burke, S., Akil, H., Watson, S., & Panksepp, J. (2003). Socially-induced brain "fertilization": Play promotes brain derived neurotrophic factor transcription in the amygdala and the dorsolateral frontal cortex in juvenile rats. *Neuroscience Letters, 341,* 17–20.

Goyal, M., Singh, S., Sibinga, E. M., Gould, N. F., Rowland-Seymour, A., Sharma, R., . . . Haythornthwaite, J. A. (2014). Meditation programs for psychological stress and well-being: A systematic review and meta-analysis. *JAMA Internal Medicine, 174*(3), 357–368.

Grabiel, A. M. (2008). Habits, ritual, and the evaluative brain. *Annual Review of Neuroscience, 31,* 359–387.

Grimm, S., Beck, J., Schuephach, D., Hell, D., Boesinger, P., Bempohl, F., . . . Northoff, G. (2008a) Imbalance between left and right dorsolateral prefrontal cortex in major depression is linked to negative emotional judgement: An fMRI study in severe major depressive disorder. *Biological Psychiatry,* 63, 369–376.

Grimm, S., Boesinger, P., Beck, J., Schuephach, D., Bempohl, F., Walter, M., . . . Northoff, G. (2008b) Altered negative BOLD responses in the default mode network during emotional processing in depressed subjects. *Neuropsychopharmacology, 34,* 932–943.

Gross-Isseroff, R., Biegon, A., Voet, H., & Wezman, A. (1998). The suicidal brain: A review of postmortem receptor/transporter binding studies. *Neuroscience and Biobehavioral Reviews, 22,* 653–661.

Guan, Z., & Fang, J. (2006). Peripheral immune activation by lipopolysaccharide decrease neurotrophins in the cortex and hippocampus in rats. *Brain, Behavior, and Immunity, 20*(1), 60–71.

Gujar, N., Yoo, S. S., Hu, P., & Walker, M. P. (2010). The unrested resting brain: Sleep deprivation alters activity within the default-mode network. *Journal of Cognitive Neuroscience, 22*(8), 1637–1648. http://doi.org/ 10.1162/ jocn.2009.21331

Harbuz, M. S., Chover-Gonzalez, A. J., & Jessop, D. S. (2003). Hypothalamic-pituitary-adrenal axis and chronic immune activation. *Annuals of the New York Academy of Science,* 99, 99–106.

Hariri, A. R., Mattay, V. S., Tessitore, A., Kolachana, B., Fera, F., Goldman, D., . . . Weinberger, D. (2002). Serotonin transporter genetic variation and the response of the human amygdala. *Science, 297,* 400–403.

Hartley, C. A., & Phelps, E. A. (2010). Changing fear: The neurocircuitry of emotion regulation. *Neuropsychopharmacology, 72,* 113–118.

Hays, S. C., & Strosahl, K. D. (2016). *Acceptance and commitment therapy: The process and practice of mindful change* (2nd ed.). New York, NY: Guilford.

Helmeke, C., Ovtscharoff, W., Jr., Poeggel, G., & Braun, K. (2008). Imbalance of immunohistochemically characterized interneuron populations in the adolescent and adult rodent medial prefrontal cortex after repeated exposure to neonatal separation stress. *Neuroscience, 152*(1), 18–28.

Heine, S. J. (2017) *DNA is not destiny.* New York: Norton.

Hickcok, G. (2014). *The myth of mirror neurons: The real neuroscience of communication and cognition.* New York, NY: Norton.

Hoch, S. L. (1998). Famine, disease, and mortality patterns in the parish of Borshevk, Russia, 1830–1912. *Population Studies: A Journal of Demography, 52*(3), 357–368.

Hoebel, B. G., Rada, P. V., Mark, G. P., & Pothos, E. N. (1999). Neural systems for reinforcement and inhibition of behavior: Relevance to eating, addiction, and depression. In D. Kahneman, E. Diener, & N. Schwarz (Eds.), *Well-being: The foundations of hedonic psychology* (pp. 558–572). New York: Russell Sage Foundation

Hofmann, S.G., Sawyer, A.T., Witt, A.A., & Oh, D. (2010). The effect of mindfulness-based therapy on anxiety and depression: A meta-analytic review. *Journal of Consulting and Clinical Psychology. 78*(2), 169-83. doi: 10.1037/a0018555

Holzel, B. K., Carmody, J., Evans, K. C., et al. (2010). Stress reduction correlates with structural changes in the amygdala. *Social Cognitive and Affective Neuroscience, 5*(1), 11–17.

Holzel, B. K., Lazar, S. W., Gard, T., Schuman-Olivier, Z., Vago, D. R., & Ott, V. (2011). How does mindfulness meditation work? Proposing mechanism of action from a conceptual and neural and neural perspective. *Perspectives on Psychological Science, 6,* 537–559.

Homberg, J. R., & Lesch, K. P. (2011). Looking on the bright side of serotonin transporter gene variation. *Biological Psychiatry, 69*(6), 513–519.

Hu, Y., Block, G., Norkus, E. P., Morrow, J. D., Dietrich, M., & Hudes, M. (2006). Relations of glycemic index and glycemic load with plasma oxidative stress markers. *American Journal of Clinical Nutrition, 84*(1), 70–76.

Iacoboni, M. (2004). Watching social interactions produces prefrontal and medial parietal BOLD fMRI signals increases compared to a resting baseline. *NeuroImage, 21,* 1167–1173.

Jablonka, E., & Lamb, M. J. (2006). *Evolution in four dimensions: Genetic, epigenetic, behavioral, and symbolic.* Cambridge, MA: MIT Press.

Jackowska, M., et al. (2012). Short sleep duration is associated with shorter telomere length in healthy men: Findings from the Whitehall II Cohort Study. *PLOS One, 7*(10), e47292. http://doi.org/10.1371/journal.pone.004292

Jacobs, T. L., Epel, E. S., Lin, J., Blackburn, E. H., Wolkowitz, O. M., Bridwell, D. A., . . . Saron, C. D. (2011). Intensive meditation training, immune cell telomerase activity, and psychological mediators. *Psychoneuroendocrinology, 36*(5), 664–681. http://doi.org/10.1016/j.psyneuen. 2010.09.010

Jaffe, J., Beebe, B., Feldstein, S., Crown, C., & Jasnow, M. (2001). Rhythms of dialogue in infancy: Coordinated timing in development. *Monographs of the Society for Research in Child Development*, 66(2), pp. i–viii, 1–149.

James, W. (1890). *Principles of Psychology*. New York, NY: Henry Holt and Company.

Janelidzes, Matte, D., Westrin, A., Traskman-Bendz, L., & Brundin, L. (2011). Cytokine levels in the blood may distinguish suicide attempters from depressed patients. *Brain, Behavior, and Immunity, 25*, 335–339.

Johnson, J. D., O'Connor, K. A., Deak, T., Stark, M., Watkins, L. R., & Maier, S. F. (2002). Prior stressor exposure sensitizing LPS-induced cytokine production. *Brain, Behavior, and Immunity, 16*, 461–476.

Johnston, E., & Olson, L. (2015). *The feeling brain: The biology and psychology of emotions*. New York, NY: Norton.

Kabat-Zinn J. (1982). An outpatient program in behavioral medicine for chronic pain patients based on the practice of mindfulness meditation: Theoretical considerations and preliminary results. *General Hospital Psychiatry, 4*, 33–47.

Kearney, D. J., McDermott, K., Malte, C., Martinez, M., Simpson, T. L. (2013). Effects of participation in a mindfulness program for veterans with posttraumatic stress disorder: A randomized controlled pilot study. *Journal of Clinical Psychology*, 69(1):14–27. doi: 10.1002/jclp.21911.

Keltikangas-Jarvinen, L., Pulkki-Raback, L., Elovainio, M., Raitakari, O. T., Viikari, J., & Lehtimaki, T. (2009). DRD2 C32806T modifies the effect of child-rearing environment on adulthood novelty seeking. *American Journal of Medical Genetics Part B: Neuropsychiatric Genetics, 150B*(3), 389–394.

Kendall-Tackett, K. (2010). Depression, hostility, posttraumatic stress disorder, and inflammation: The corrosive health effects of negative mental states. In K. Kendall-Tackett (Ed.), *The psychoneuroimmunology of chronic disease* (pp. 113–132). Washington: American Psychological Association.

Kendler, K. S. (2005). "A gene for . . . " The nature of gene action in psychiatric disorders. *American Journal of Psychiatry, 162*, 124–125.

Kessler, R. C., Ormel, J., Demler, O., & Stang, P. E. (2003). Comorbid mental disorders account for the role impairment of commonly occurring chronic physical disorders: Results from the National Comorbidity Survey. *Journal of Occupational and Environmental Medicine, 45*, 1257–1266. http://doi.org/10.1097/01

Kiecolt-Glaser, J. K., Garner, W., Speicher, C., Penn, G. M., Holliday, J., & Glaser, R. (1984). Psychosocial modifiers of immunocompetence in medical students. *Psychosomatic Medicine, 46*(1), 7–14.

Kilpatrick, D. G., Koenen, K. C., Ruggiero, K. J., et al. (2007). The serotonin transporter gene and social support and moderation of posttraumatic stress disorder and depression in hurricane-exposed adults. *American Journal of Psychiatry, 164*, 1693–1699.

Konner, M., & Worthman, C. (1980). Nursing frequency, gonadal function and birth spacing among the Kung hunter-gatherers. *Science, 207*, 788–791.

Kok, B. E., Coffey, K. A., Cohn, M. A., Catalino, L. I., Vacharkulksemsuk, T. Algoe, S. B., Brantley, M. &. Fredrickson. B. L. (2013). How positive emotions build physical health: Perceived positive social connections account for the upward spiral between positive emotions and vagal tone. *Psychological Science, 24*(7), 1123–1132.

Koyama, T., McHaffie, J. G., Laurienti, P. J., & Coghill, R. C. (2005). The subjective experience of pain: Where expectations become reality. *Proceedings of the National Academy of Sciences of the USA, 102*(36), 12950–12955.

Kradin, R. (2008). *The placebo response and the power of unconscious healing.* New York, NY: Routledge.

Kumsta, R., & Heinrichs, M. (2013). Oxytocin, stress, and social behavior: Neurogenetics of the human oxytocin system. *Current Opinion in Neurobiology, 2*, 11–16.

Lambert. K. (2008). *Lifting depression: A neuroscientist's hands-on approach to activating your brain's healing power.* New York: Basic Books.

Lanius, R. A., Williamson, P. C., Bluhm, R. L., Densmore, M., Boksman, K., Neufeld, R. W. J., . . . Menon, R. S. (2005). Functional connectivity of dissociative responses in posttraumatic stress disorder: A functional magnetic resonance imaging investigation. *Biological Psychiatry, 57*(8), 873–884.

Lazar, S. W., Kerr, C. E., Wasserman, R. H., Gray, J. R., Greve, D. N., Treadway, M. T., McGarvey, M., Quinn, B. T., Dusek, J. A., Benson, H., Rauch, S. L., Moore, C. I., Fischl, B. (2005). Meditation experience is associated with increased cortical thickness. *Neuroreport, 16*, 1893–1897.

LeDoux, J. (2016). *Anxious: Using the brain to understand and treat fear and anxiety.* New York, NY: Penguin.

Lee, J. Y., et al. (2015). Association between dietary patterns in the remote past and telomere length. *European Journal of Clinical Nutrition*, 69(9), 1048–1052. http://doi.org/10.1038/ejcn.2015.58

Lee, R., Geracioti, T. D., Kasckow, J. W., & Coccaro, E. F. (2005). Childhood trauma and personality disorder: Positive correlation with adult CSF

Corticotropin-Releasing Factor Concentrations, *American Journal of Psychiatry, 162*, 995-997.

Lepore, S. J., Miles, H. J., & Levy, J. S. (1997). Relation of chronic and episodic stressors to psychological distress, reactivity, and health problems. *International Journal of Behavioral Medicine, 4*, 39–59.

Levine, P. (2015). *Trauma and memory: brain and body in a search for the living past: A practical guide for understanding and working with traumatic memory*. New York: North Atlantic Books.

Lieberman, M. D. (2013). *Social: Why our brains are wired to connect*. New York: Crown.

Lim, A., & Marsland, A. (2014). Peripheral pro-inflammatory cytokines and cognitive aging: The role of metabolic risk. In A.W. Kusnecov and H. Anisman (Eds.), *The Wiley-Blackwell Handbook of Psychoneuroimmunology*, Chapter 17, 330–346.

Lin, J., Epel, E. S., & Blackburn, E. H. (2009). Telomere, maintenance, and the aging of cells and organisms. In G. G. Berntson & J. T. Cacioppo (Eds.), *Handbook of neuroscience for the behavioral sciences* (Vol. 2, pp. 1280–1296). Hoboken, NJ: Wiley.

Linehan, M. (1993). *Cognitive-Behavioral Treatment of Borderline Personality Disorder*. New York: Guilford Press

Liu, S., Manson, J.E., Buring, J.E., Stampfer, M.J., Willett, W.C., Ridker, P.M. (2002). Relation between a diet with a high glycemic load and plasma concentrations of high-sensitivity C-reactive protein in middle-aged women. *American Journal of Clinical Nutrition, 75*(3), 492–498.

Liu, Z., Xu, C., Xu, Y., et al. (2010). Decreased regional homogeneity in insula and cerebellum: A resting-state fMRI study in patients with major depression and subjects at high risk for major depression. *Psychiatry Research, 182*, 221–215.

Lockwood, L. E., Su, S., & Youssef, N. A. (2015). The role of epigenetics in depression and suicide: A platform for gene-environment interactions. *Psychiatry Research, 228*, 235–242.

Lutz, A., Slagter, H. A., Dunne, J. D., & Davidson, R. J. (2008). Attention regulation and monitoring in meditation. *Trends in Cognitive Science, 12*, 163–169.

Ma, S. H., & Teasdale, J. (2004). Mindfulness-based cognitive therapy for depression: Replication and exploration of differential relapse prevention effects. *Journal of Consulting and Clinical Psychology, 72*(1), 31–40.

Maes, M., Kubera, M., Leunis, J.C. (2008). The gut-brain barrier in major depression: Intestinal mucosal dysfunction with an increased translocation of LPS from gram-negative enterobacteria (leaky gut) plays a role in inflammatory pathophysiology of depression. *Neuroendocrinology Letters, 29*(1), 117–124.

Maier, S. F., & Watkins, L. R. (2009). Neuroimmunology. In G. G. Berntson & J. T. Cacioppo (Eds.), *Handbook of neuroscience for the behavioral sciences* (Vol. 1, pp. 119–135). Hoboken: Wiley.

Major Depressive Disorder Working Group, Psychiatric GWAS Consortium. (2013). A mega-analysis of genome-wide association studies for major depressive disorder. *Molecular Psychiatry, 18*(4), 497–511.

Maletic, V., & Raison, C. (2017). *The new mind-body science of depression.* New York, NY: Norton.

Mamdani, F., Rollins, B., Morgan, L., Myers, R.M., Barchas, J.D., Schatzberg, A.F., Watson. S.J., . . . Sequeira, P.A. (2015). Variable telomere length across post-mortem human brain regions and specific reductions in the hippocampus of major depressive disorder. *Translational Psychiatry, 5,* e636. http://doi.org/10.1038/tp.2015.134

Martikainen, I. K., Nuechterline, E. B., Pecina, M., Love, T. M., Cumminford, C. M., Green C. R., . . . Zubieta, J. K. (2015). Chronic back pain is associated with alterations in dopamine neurotransmission in the ventral striatum. *Journal of Neuroscience, 35*(27), 9957–9965.

Masten, C. L., Morelli, S. A., & Eisenberger, N. I. (2011). An fMRI investigation of empathy for "social pain" and subsequent prosocial behavior. *NeuroImage, 55,* 381–388.

Mayberg, H. J., Liotti, M., Brannan, S. K., et al. (1999). Reciprocal limbic-cortical function and negative mood: Converging PET findings in depression and normal sadness. *American Journal of Psychiatry, 156,* 675–682.

Mayberg, H. J., Lozano, A., Voon, V., et al. (2005). Deep brain stimulation for treatment-resistant depression. *Neuron, 45*(5), 651–660.

Mayberg, H. J., Silva, J. A., Brannan, S. K., Tekell, J. L., Mahurin, R. K., McGinnis, S., et al. (2002). The functional neuroanatomy of the placebo response. *American Journal of Psychiatry, 159,* 728–737.

Mayer, E. (2016). *The mind-gut connection: How the hidden conversation within our bodies impact our mood, our choices, and overall health.* New York, NY: HarperWave.

McCullough, M. E., Emmons, R. A., & Tsang, J. (2002). The grateful disposition: A conceptual and empirical topography. *Journal of Personality and Social Psychology, 82,* 112–127.

McDade, T. W., Hawkley, L. C., & Cacioppo, J. T. (2006). Psychosocial and behavioral predictors of inflammation in middle-aged and older adults: The Chicago Health, Aging, and Social Relations Study. *Psychosomatic Medicine, 68,* 517–523.

McEwen, B., & Morrison, J. (2013). Brain on stress: Vulnerability and plasticity of the prefrontal cortex over the course. *Neuron, 79,* 16–29. http://doi.org/10.1016/j.neuron.2013.06.028

McEwen, B., & Wingfield, J. (2003). The concept of allostasis in biology biomedicine. *Hormone and Behaviors, 43,* 2–15.

McCraty, R., & Childre, D. (2004). The grateful heart: The psychophysiology of appreciation. In R. A. Emmons & M. E. McCullough (Eds.), *The psychology of gratitude* (pp. 230-255). New York: Oxford University Press.

McEwen, B.S. (2002). *The end of stress as we know it.* Washington, D. C: Joseph Henry Press

McFadden, K. L., Marc-Andre Cornier, M., Edward, L., Melanson, E. L., Jamie, L., Bechtell, J. L., & Tregellas, J. R. (2013). Effects of exercise on resting-state default mode and salience network activity in overweight/obese adults. *Neuroreport, 24*(15), 866–871.

McGowan, P. O., Aya Sasaki, A., D'Alessio, A. C., Dymov, S., Labonté, B., Szyf, M., . . . Meaney, M. J. (2009). Epigenetic regulation of the glucocorticoid receptor in human brain associates with childhood abuse. *Nature Neuroscience, 12,* 342–348.

McGraty, R., & Atkinson, M. (2004). The grateful heart: The psychophysiology of appreciation. In R. A. Emmons & M. E. McCullough (Eds.), *The psychophysiology of gratitude* (pp. 230–255). New York, NY: Oxford University Press.

Milad, M. R., & Quirk, G. J. (2012). Fear extinction as a model for translation neuroscience: Ten years of progress. *Annual Review of Psychology, 63,* 129–151.

Miller, G. E., Chen, E., Sze, J., Marin, T., Arevalo, J. M., Doll, R., et al. (2008). A functional genomic fingerprint of chronic stress in humans: Blunted glucocorticoid and increased NF-kappaB signaling. *Biological Psychiatry, 64,* 266–272.

Miller, W., & Rollnick, S. (2012). *Motivational interviewing: Helping people change* (3rd ed.). New York, NY: Guilford.

Mitchell, J. P., Banaji, M. R., & Macrae, C. N. (2005). The link between social cognition and self-referential thought in the medial prefrontal cortex. *Journal of Cognitive Neuroscience, 17,* 1306–1315.

Moerman, D. (2002). *Meaning, medicine, and the placebo effect.* Cambridge, UK: Cambridge University Press.

Moll, J. Krueger F, Zahn R, Pardini M, de Oliveira-Souza R, Grafman, J. (2006). *Human fronto-mesolimbic networks guide decisions about charitable donation. Proceedings of the National Academy of Science, 103,* 15623–15628.

Morgan, N., Irwin, M. R., Chung, M., & Wang, C. (2014). The effects of mind-body therapies on the immune system: Meta-analysis. *PLoS One*, 9(7), e100903. https://doi.org/10.1371/journal.pone.0100903

Moriceau, S., Shionoya, K., Jakubs, K., & Sullivan, R. M. (2009). Early-life stress disrupts attachment learning: The role of amygdala corticosterone, locus ceruleus corticotropin releasing hormone, and olfactory bulb norepinephrine. *Journal of Neuroscience, 29,* 15745–15755.

Morris, R. (2007). Theories of hippocampal function. In P. Andersen, R. Morris, D. Amaral, T. Bliss, & J. O'Keefe (Eds.), *The hippocampus book* (pp. 581–714). New York, NY: Oxford University Press.

Moussavi, S., Chatterji, S., Verdes, E., Tandon, A., Patel, V., & Ustun, B. (2007). Depression, chronic diseases, and decrements, in health: Results from the World Health Surveys. *Lancet, 370*(9590), 851–858.

Muller-Pinzler, L., Krach, S., Kramer, U. M., & Paulus, F. M. (2016). The social neuroscience of interpersonal emotions. *Current Topics in Behavioral Neuroscience.* http://doi.org/ 10.1007/7854_2016437

Nanni, V., Uher, R., & Danese, A. (2011). Childhood maltreatment predicts unfavorable course of illness and treatment outcome in depression: A meta-analysis. *American Journal of Psychiatry, 169*(2), 141–151.

Neff, K. D. (2011). Self-compassion, self-esteem, and well-being, *Social and Personality Psychology Compass*, 5(1), 1–12.

Neeper, S. A., Gomez-Pinilla, F., Choi, J., & Cotman, C. W. (1996). Physical activity increases mRNA for brain-derived neurotrophic factor and nerve growth factor in the rat brain. *Brain Research, 726,* 49–56.

Neuman, I. D. (2008). Brain oxytocin: A key regulator of emotional and social behaviors in both females and males. *Journal of Neuroendocrinology, 20,* 858–865.

Newcomber, J., Selke, G., Melson, Gross, J., Bogler, G., & Dagogo-Jack, S. (1998). Dose-dependent cortisol-induced increases in plasma leptin concentration in healthy humans. *Archives of General Psychiatry, 55,* 995–1000.

Nezu, A. M., Raggio, G., Evans, A. N., & Nezu, C. H. (2013). Diabetes mellitus. In A. M. Nezu, C. H. Nezu, & P. A. Geller (Eds.), *Handbook of psychology: Health psychology* (Vol. 9, pp. 200–217). Hoboken, NJ: Wiley.

Nixon, N. L., Liddle, P. F., Nixon, E., Worwood, G., Liotti, M., & Palaniyappan, L. (2014). Biological vulnerability to depression: Linked structural and functional brain network findings. *British Journal of Psychiatry*, 204, 283–289.

Northoff, G., & Bempohl, F. (2004). Cortical midline structures and the self. *Trends in Cognitive Sciences*, 8, 102–107.

Ogden, P., & Fischer, J. (2015). *Sensorimotor psychotherapy.* New York, NY: Norton.

O'Hanlon, B., & Rowan, T. (2003). *Solution oriented therapy*. New York, NY: Norton.

Osorio, R. S., Gumb, T., Pirraglia, E., Varga, A. W., Lu, S. E., Lim, J., . . . Alzheimer's Disease Neuroimaging Initiative. (2015). Sleep-disordered breathing advances cognitive decline in the elderly. *Neuroscience, 282C*, 109–121.

Panksepp, J., & Biven, L. (2012). *The archaeology of mind*. New York: Norton.

Parker, K. J., Buckmaster, C. L., Lindley, S. E., Schatzberg, A. F., & Lyons, D. M. (2012). Hypothalamic-pituitary-adrenal axis physiology and cognitive control of behavior in stress inoculated monkeys. *International Journal of Behavioral Development, 36*(1), 10.1177/0165025411406864. http://doi.org/10.1177/0165025411406864

Paquette, D. & Bigras, M. (2010) The risky situation: a procedure for assessing the father–child activation relationship, *Early Child Development and Care, 180*(1–2), 33-50.

Pect, M., & Stokes, C. (2005). Omega-3 fatty acids in the treatment of psychiatric disorders. *Drugs, 65*, 1051–1059.

Petry, N. M. (2000). A comprehensive guide to the application of contingency management procedures in clinical settings. *Drug and Alcohol Dependence, 58*, 9–25.

Pezawas, L., Meyer-Lindenberg, A., Drabant, E. M., et al. (2005). 5HTTLPR polymorphism impacts human cingulate-amygdala interactions: A genetic susceptibility mechanism for depression. *Nature Neuroscience, 8*, 828–834.

Phelps, E. (2009). The human amygdala and the control of fear. In P. J. Whalen & E. A. Phelps (Eds.), *The human amygdala* (pp. 204–219). New York, NY: Guilford.

Porges, S. (2011). *The polyvagal theory: Neurophysiological foundations of emotion, attachment, communication, and self-regulation*. New York, NY: Norton.

Posner, J., Hellerstein, D. J., Gat, I., Mechling, A., Klahr, K., Wang, Z., . . . Bradley, S. P. (2013). Antidepressants normalize the default mode network in patients with dysthymia. *JAMA Psychiatry, 70*(4), 373–382. http://doi.org/10.1001/jamapsychiatry.2013.455.

Pratico, D., Christopher, M., Clark, C. M., Liun, F., Rokach, J., Lee, V., & Trojanowski, J. Q. (2002). Increase of brain oxidative stress in mild cognitive impairment: A possible predictor of Alzheimer disease. *Archives of Neurology, 59*(6), 972–976.

Raichle, M.E., (2006). The brain's dark energy. *Science, 314*(5803), 1249–1250.

Raison, C.L., Capuron, L., Miller, A.H. (2006). Cytokines sing the blues: Inflammation and the pathogenesis of depression. *Trends in Immunology, 27*, 24–31.

Raison, C. L., & Miller, A. H. (2001). The neuroimmunology of stress and depression. *Seminars in Clinical Neuropsychiatry*, 6(4), 277–294.

Raphael, K. G., Janal, M. N., Nayak, S., Schwartz, J. E., & Gallagher, R. M. (2004). Familial aggregation of depression in fibromyalgia: A community-based test of alternate hypotheses. *Pain, 110*(1–2), 449–460.

Resnick, H., & Howard, B. (2002). Diabetes and cardiovascular disease. *Annual Review of Medicine, 53*, 245–267.

Reul, L. M. H., Collins, A., & Gutierrez-Mecinas, M. (2011). Stress effects on the brain: Intracellular signaling cascades, epigenetic mechanisms, implications in behavior. In C. D. Conrad (Ed.), *Handbook of stress: Neuropsychological effects on the brain* (pp. 95–113). Hoboken: Wiley-Blackwell.

Rizzolatti G, & Arbib M.A. (1998) Language within our grasp. *Trends in Neuroscience, 21*(5), 188–194.

Rohr, R. (2015). *What mystics know*. Chestnut Ridge, NY: Crossroad.

Rominger, A., Cumming, P., Xiong, G., Koller, G., Boning, G., Wulff, M., . . . Pogarell, O. (2012). [18F]Fallypride PET measurement of striatal and extrastriatal dopamine D2/3 receptor available in recently abstinent alcoholics. *Addiction Biology, 17*(2), 490–503.

Rossi, E. (2002). *The psychobiology of gene expression: Neuroscience and neurogenesis in hypnosis and the healing arts*. New York: Norton.

Russell, D. W., & Cutrona, C. E. (1991). Social support, stress, and depressive symptoms among the elderly: test of a process model. *Psychological Aging*, 6(2), 190–201.

Sanchez-Villegas, A., Delgado-Rodriguez, M., Alonso, A., Schlatter, J., Lahortiga, F., Serra Majem, L., Martinez-Gonzalez, M.A. (2009). Association of the Mediterranean dietary pattern with the incidence of depression: The Seguimiento Universidad de Navarra/University of Navarra Follow-up (SUN) cohort. *Archives of General Psychiatry*, 66(10), 1090–1098. http://doi.org/10.100/archgenpsychiatry.2009.129

Sapirstein, G., & Kirsch, I. (1988). Listening to Prozac but hearing placebo? A meta-analysis of the placebo effect of antidepressant medication. *Prevention and Treatment, 1*, 3–11.

Sareen, J., Cox, B. J., Stein, M. B., Afifi, T. O., Fleet, C., & Asmundson,G.J. (2007). Physical and mental comorbidity, disability, and suicidal behavior associated with posttraumatic stress disorder in a large community sample. *Psychosomatic Medicine*, 69(3), 242–248.

Schedlowski, M., Jacobs, R., Alker, J., Prohl, F., Stratmann, G., Richter, S., et al. (1993). Psychophysiological, neuroendrocrine, and cellular immune reactions under psychological stress. *Neuropsychobiology, 28*, 87–90.

Schienle, A., Schafer, A., Hermann, A., Rohrmann, S., & Vaitl, D. (2007). Symptom provocation and reduction in patients suffering from spider phobia: An fMRI study on exposure therapy. *European Archives of Psychiatry Clinical Neuroscience, 257*(8), 486–493.

Schore, A. (1999). *Affect regulation and the origin of the self: The neurobiology of emotional development.* New York, NY: Norton.

Scott, D. J., Stohler, C. S., Egnatuk, C. M., Wang, H., Koeppe, R. A., & Zubieta, J. K. (2007). Individual differences in reward responding explain placebo-induced expectations and effects. *Neuron, 55*(2), 25–36.

Seckl, J. R. (2008). Glococorticoids, developmental programming and the risk of affective dysfunction. *Progress in Brain Research, 167,* 17–34.

Sharot, T., Riccardi, A. M., Raio, C. M., & Phelps, E. A. (2007). Neural basis of egalitarian bias. *Nature, 450,* 102–105.

Sharpley, C. F., Palanisamy, S. K., Glyde, N. S., Dillingham, P. W., & Agnew, L. L. (2014). An update on the interaction between the serotonin transporter promoter variant (5-HTTLPR), stress and depression, plus an exploration of non-conforming findings. *Behavioral Brain Research, 273,* 89–105.

Shiota, M. N., Campos, B., Oveis, C., Hertenstein, M. J., Simon-Thomas, E., & Keltner, D. (2017). Beyond happiness: Building a science of discrete positive emotions. *American Psychologist, 72*(7), 617–643.

Siegel, D. (2015). *The developing mind, second edition: How relationships and the brain interact to shape who we are.* New York: Guilford Press.

Siegel, D. J. (2017). *Mind: A journey to the heart of being human.* New York, NY: Norton.

Singer, T., Seymore, B., O'Doherty, J., Kaube, H., Dolan, R. J., & Firth, C. D. (2014). Empathy for pain involves the affective but not the sensory components of pain. *Science, 303,* 1157–1161.

Slagter, H. A., Lutz, A., Gresichar, L. L., Francis, A. D., Nieuwenhuis, S., Davis, J. M., . . . Cole, S. W. (2007). Social stress enhances sympathetic innervation of primate lymph nodes: Mechanisms and implications for viral pathogenesis. *Journal of Neuroscience, 27,* 8857–8865.

Sokolov, E., Spinks, J., Naatanen, R., & Heikki, L. (2002). *The orienting response in information processing.* Mahwah, NJ: Erlbaum.

Sokolov, E. N. (1990). The orienting response, and future directions of its development. *Pavlovian Journal of Biological Science. 25*(3), 142–50.

Solomon, G. F. (1987). Update on psychoneuroimmunology. *Western Journal of Medicine, 147*(1), 72.

Spitzer, S. B., Llabre, M. M., Ironson, G. H., Gellman, M. D., Schneiderman,

N. (1992). The influence of social situations on ambulatory blood pressure. *Psychosomatic Medicine, 54,* 79–86.

Stein, A. D., Ravelli, A. C., et al. (1995). Famine, third-trimester pregnancy weight gain, and intrauterine growth: The Dutch Famine Birth Cohort Study. *Human Biology, 67*(1), 135–150.

Stellar, J. E., Cohen. A., Oveis, C., & Keltner, D. (2015) Affective and physiological responses to the suffering of others: compassion and vagal activity. *Journal of Personality and Social Psychology, 108*(4), 572-585. doi: 10.1037/pspi0000010.

Steptoe A, Hamer M, & Chida Y. (2007) The effects of acute psychological stress on circulating inflammatory factors in humans: A review and meta-analysis. Brain, Behavior, and Immunity;21:901–912. doi: 10.1016/j.bbi.2007.03.011.

Sterling, P., & Eyer (1988). Allostasis: A new paradigm to explain arousal pathology. In S. Fisher & J. Reason (Eds.), *Handbook of life stress, cognition, and health* (pp. 629–649). New York, NY: Wiley.

Steward, J. (2006). The detrimental effects of allostasis: Allostatic load as a measure of cumulative stress. *Journal of Physiological Anthropology, 25,* 133–145.

Stieglitz, J., Trumble, B. C. Thompson, M. E., Blackwell, A. D., Kaplan, H., & Gurven, M. (2015). Depression as sickness behavior? A test of the host defense hypothesis in a high pathogen population. *Brain, Behavior, and Immunity, 49,* 130–139.

Stickgold, R. (2002). EMDR: a putative neurobiological mechanism of action. *Journal of Clinical Psychology, 58*(1), 61–75.

Strickgold, R. (2015, October). Sleep on it! Your nightly rest turns out to affect your mind and health more than anyone suspected. *Scientific American,* 32–57.

Suarez, E. C., Lewis, J. G., Krishnan, R. R., & Young, K. H. (2004). Enhanced expression of cytokines and chemokines by blood monocytes to in vitro lipopolysaccharide stimulation are associated with hostility and severity of depressive symptoms in healthy women. *Psychoneuroimmunology, 29,* 1119–1128.

Swain, R. A., Harris, A. B., Wiener, E. C., Dutka, M. V., Morris, H. D., Theien, B. E., et al. (2003). Prolong exercise induces angiogenesis and increases cerebral blood volume in primary cortex of the rat. *Neurogenesis, 117,* 1037–1046

Szyf, M. (2011). The early life social environment and DNA methylation: DNA methylation mediating the long-term impact of social environments in early life. *Epigenetics, 6,* 971–978.

Tang, Y. Y., Lu, Q., Geng, X., Stein, F. A., Yang, Y., & Posner, M. I. (2010). Short-term meditation induces white matter changes in the interior cingulate. *Proceedings of the National Academy of Sciences of the USA, 107,* 15649–15652.

Taylor, V. A., Daneault, V., Grant, J., et al. (2013). Impact of meditation training on the default mode network during a restful state. *Social Cognitive and Affective Neuroscience*, 8(1), 4–14.

Teasdale, J., Segal, Z. V., Williams, J. M., et al. (2000). Prevention of relapse/recurrence in major depression by mindfulness-based cognitive therapy. *Journal of Consulting and Clinical Psychology*, 68(4), 31–40.

Tedeshi, R.G.; Calhoun, L.G. (1996). The posttraumatic growth inventory: Measuring the positive legacy of trauma. *Journal of Traumatic Stress*, 9, 455–471.

Teicher, M. H., Samson, J. A., Polcari, A., & McGreenery, C. E. (2006). Sticks, stones, and hurtful words: Relative effects of various forms of childhood maltreatment. *American Journal of Psychiatry, 163*(6), 993–1000.

Tonelli, M.E., & Wachholtz, A.B. (2014). Meditation-based treatment yielding immediate relief for meditation-naïve migraineurs. *Pain Management Nursing, 15*(1), 36–40. doi: 10.1016/j.pmn.2012.04.002. Epub 2012 Jun 20.

Tottenham, N. (2014). The importance of early experiences for neuro-affective development. *Current Topics in Behavioral Neuroscience, 16*, 109–129. http://doi.org/10.1007/7854_254

Tottenham, N., Hare, T. A., & Casey, B. J. (2009). A developmental perspective on human amygdala function. In P. J. Whalen & E. A. Phelps (Eds.), *The human amygdala* (pp. 107–117). New York, NY: Guilford.

Tracey, K. J. (2002). The inflammatory reflex. *Nature, 420*(6917), 857.

Tracey, K. J. (2009). Reflex control of immunity. *Nature Reviews Immunology*, 9(6), 418–428. http://doi.org/10.1038/nri2566

Trafton, J. A., & Gifford, E. V. (2008). Behavioral reactivity and addiction: The adaptation of behavioral response to reward opportunities. *Journal of Neuropsychiatry and Clinical Neuroscience, 20*(1), 23–35.

Trafton, J. A., Gordon, W., Misra, S. (2016). *Training your brain to adopt healthful habits: Mastering the five brain challenges*, Second Edition. Los Altos, CA: Institute for Brain Potential.

Tronick, E. (2007). *The neurobehavioral and social-emotional development of infants and children*. New York, NY: Norton.

van der Kolk, B. (2015). *The body knows the score*. New York, NY: Penguin.

van IJzendoorn, M. H., & Bakermans-Kranenburg, M. J. (1996). Attachment representations in mothers, fathers, adolescents and clinical groups: A meta-analytic search for normative data. *Journal of Consulting and Clinical Psychology*, 64, 8–21.

Vanuytsel, T., et al. (2014). Psychological stress and corticotropin-releasing hormone increase intestinal permeability in humans by mast cells-dependent mechanism. *Gut*, 63(8), 1293–1299. http://doi.org/10.1136/gutjnl-2013-305690

Venkatraman, V., Chuah, Y. M., Huettel, S. A., & Chee, M. W. (2007). Sleep deprivation elevates expectation of gains and attenuates response to losses following risky decisions. *Sleep, 30*(5), 603–609.

Verhoeven, J. E., et al. (2014). Major depressive disorder and accelerated cellular aging: Results from a large psychiatric cohort study. *Molecular Psychiatry, 19*(8), 895–901. http://doi.org/10.1038/mp.2013.151

Vgontzas, A. N., Zoumakis, M., Papanicolaou, D. A., Bixler, E. O., Prolo, P., Lin, H. M., et al. (2002). Chronic insomnia is associated with a shift of interleukin-6 and tu mor necrosis factor secretion from nighttime to daytime. *Metabolism 51*, 887–892.

Vogt, B. A. & Sikes, R. W. (2009). Cingulate nociceptive circuitry and roles in pain processing: the cingulate premotor pain model. In B. V. Vogt (Ed.) *Cingulate neurobiology and disease* (pp. 311–338) New York, NY: Oxford University Press.

Vytal, K., & Hamann, S. (2010). Neuroimaging support for discrete neural correlates of basic emotions: A voxel-based meta-analysis. *Journal of Cognitive Neuroscience, 22*, 2864–2885.

Uher, R. (2011). Genes, environment, and individual differences in responding to treatment for depression. *Harvard Review of Psychiatry, 19*, 109–124.

Wager, T. D., Davidson, M. L., Hughes, B. L., Lindquist, M. A., & Ochsner, K. N. (2008). Prefrontal-subcortical pathways mediating successful emotional regulation. *Neuron, 59*, 1037–1050.

Wager, T. D., Rilling, J. K., Smith, E. E., Sokolik, A., Casey, K. L., Davidson, R. J., . . . Cohen, J. D. (2004). Placebo-induced changes in fMRI in the anticipation and experience of pain. *Science, 303*(5661), 1162–1167.

Weaver, I. C., Cervoni, N., et al. (2004). Epigenetic programing by maternal behavior. *Nature Neuroscience, 7*(8), 847–854.

Weinberg, A., & Hajcak, G. (2011). Longer-term test-retest reliability of error-related brain activity. *Psychophysiology*, 48, 1420–1425.

Wen. D. J., Soe, N. N., Sanmugam, S., Kwek, K., Chong, Y. S., Gluckman, P. D., Meaney, M. J., Rifkin-Graboi, A, & Qui, A. (2017). Infant frontal EEG asymmetry in relation with postnatal maternal depression and parenting behavior. *Translational Psychiatry, 7*, 1057.

Weng, H. Y., Fox, A. S., Shackman, A. J., Stodola, D. E., Caldwell, J. Z., Olson, M. C., Rogers, G. M., & Davidson, R. J. (2013). Compassion training alters altruism and neural responses to suffering. *Psycholological Science, 24*(7), 1171–1180. doi: 10.1177/0956797612469537. Epub 2013 May 21.

Whalen, P. J. & Phelps E. A. (2009). *The human amygdala*. New York, NY: Guilford.

Whitmer, R. A., Gunderson, E. P., Barrett-Connor, E., Quesenberry, C. P., & Yaffe, K. (2005). Obesity in middle age and the future risk of dementia: A 27 year longitudinal population based study. *British Medical Journal, 330,* 1360–1362.

Whitmer, R. A., Gustafson, D. R., Barrett-Connor, E., Haan, M. N., & Yaffe, K. (2008). Central obesity and increased risk of dementia more than three decades later. *Neurology, 71,* 1057–1064.

Wojczynski, M. K., North, K. E., Pedersen, N. L., & Sullivan, P. F. (2007). Irritable bowel syndrome: A co-twin control analysis. *American Journal of Gastroenterology, 102*(10), 2200–2229.

Xie, L., Kang, H., Xu, Q., Chen, M. J., Liao, Y., Thiyagarajan, M., O'Donnell, J., Christensen, D., Nicholson, C., Iliff., J., Takano, T. Deane, R., & Nedergaard, M. (2013). Sleep drives metabolite clearance from the adult brain. *Science, 342*(6156), http://doi.org/10.1126/science.1241224

Yehuda, R. A., & Bierer, L. (2007). Transgenerational transmission of cortisol and PTSD risk. *Progress in Brain Research, 167,* 121–135.

Yehuda, R. A. Engel, S. M., et al. (2005). Transgenerational effects of posttraumatic stress disorder in babies of mothers exposed to the World Trade Center attacks during pregnancy. *Journal of Clinical Endocrinology and Metabolism,* 90(7), 4115–4118.

Zak, P. J. (2012). *The moral molecule: How trust works.* New York, NY: Penguin.

Zak, P. J., Stanton, A. A., & Ahmadi, S. (2007). Oxytocin increases generosity in humans. *PLoS One, 2,* e11228.

Zhang, R., et al. (2009). Circulating endotoxin and systemic immune activation in sporadic amyotrophic lateral sclerosis (sALS). *Journal of Neuroimmunology, 206*(1–2), 121–124. http://doi.org/10.1016/j.nuroim.2008.09.017

Zhang, T. Y., & Meaney, M. J. (2010). Epigenetics and the environmental regulation of the genome and its function. *Annual Review of Psychology, 61,* 439–466.

Index

AAI. *see* Adult Attachment Interview (AAI)
A1 allele of D2 dopamine receptor, 140–41
  alcohol abuse related to, 141
  impulsive behaviors related to, 141
absorption, 209–10
abuse
  inflammation due to, 65
  substance, 76–78
ACC. *see* anterior cingulate cortex (ACC)
accelerated aging
  shortening of telomeres and, 87
acceptance and commitment therapy, xxx, 265–66
  for anxiety disorders, 266
ACEs. *see* adverse childhood experiences (ACEs)
ACE study. *see* Adverse Childhood Experiences (ACE) study
acetyl
  function of, 72
acetylation, 72
  gene suppression through, xxiii
acetylcholine
  receptors for, 165–66
acid(s)
  amino, 123
  docosahexaenoic, 122
  eicosapentaenoic, 122
  fatty *see* fatty acids
acquired characteristics
  theory of inheritance of, 68
ACTH. *see* adrenocorticotropic hormone (ACTH)
acting "as if," 189
action-oriented empathy, 254–55
active listening
  in motivational interviewing, 57
activity
  coactivation of mental operating systems effects on, 260
  synchronized, 264
acute inflammation
  chronic inflammation *vs.*, 102
acute stress
  case example, 172–75
  depression and, 199
  progressive effects of, 172–75
acute stress disorders
  PTSD from, 134
acute stressor(s)
  immune responses in, 168
adaptation
  early patterns of, 42
  sickness behavior in, 225
addiction(s), 139–61
  in ACC activation, 157
  accessing reward circuits in working with, 160
  anxiety and, xxvi
  building of strengths related to, 156–57
  dopamine in, 142–43, 144f
  drug delivery methods effects on, 154–55
  drugs and, 154–55
  gambling and, 142–45, 144f
  habits becoming, xxvi
  motivation and, 154–59, 158f

addiction(s) (*continued*)
motivation–pleasure circuit and, 141–46
nondrug, 155
predilection for, 156
repeated use and, 155
ADH. *see* alcohol dehydrogenase (ADH)
adrenaline, 172
adrenocorticotropic hormone (ACTH), 80, 166
release of, 179
adult attachment
variations in, 47–49
Adult Attachment Interview (AAI), 48
adult attachment styles
assessment of, 48
advanced glycation end products (AGEs)
adverse effects of, 118–19
adverse childhood experiences (ACEs)
anxiety related to, 79
case examples, 195
categories of, 62
depression related to, 79, 219–20, 223
destructive habits associated with, 63–64
DMN effects of, 14
dopamine and cortisol levels related to, 141
epigenetic factors enhancing effects of nurturance or exacerbating effects of, 75
genetic vulnerability and, 74
health consequences of, 61–64
poor self-care associated with, 63–64
public burden of, 64
short version of serotonin transporter gene and, 74–75
suicide related to, 79
adverse childhood experiences (ACEs) questions
in health surveys, 64
Adverse Childhood Experiences (ACE) study
described, 62–64
health and mental health ramifications of, xxii–xxiii
Adverse Childhood Experiences (ACE) study scores, 62–63
autoimmune disorders later in adult life related to, 65
correlations with adverse outcomes, 63–64
depression and suicide related to, 79
aerobic exercise
benefits of, xxv
affective systems, 3
affect regulatory problems
prefrontal deficits related to, 147
affect stability
early deprivation impact on, xxii
AGEs. *see* advanced glycation end products (AGEs)
aging
accelerated, 87
inflammation related to, 126
Akkermansia
BMI related to, 111
in lean people, 111
alcohol
as reward and relief from stress, 154
sleep suppressed by, 136–37
alcohol abuse
A1 allele of D2 dopamine receptor and, 141
alcohol dehydrogenase (ADH)
enzyme activity associated with, 77–78
Alcoholics Anonymous, 148, 151, 257
alcoholism
case example, 154
cues in, 154
reward-seeking responses in, 154
alcohol use
cannabis smoking *vs.*, 77–78
case example, 77–78
alkalosis
hypocapnic, 187
allostasis, xxvi–xvii, 168–72
defined, 169
described, 170
mediators of, 169–70
allostatic load, xxvi–xvii, 168–72
adverse effects of, 171–72
case example, 172–75
cumulative effects of, 173
described, xxvii, 171
Alzheimer's disease
as "diabetes type 3," 102
exercise in lowering risk of, 124
isoprostanes and, 119
LPS in, 111
obesity and, 101

INDEX

ambiguity
  intolerance to, 183–84
  learning to tolerate, 184
amino acids
  as building blocks for neurotransmitters, 123
amygdala, 160
  cortisol effects on, 174
  excessive activity of, 229–30
  in exposure to stress-provoking stimuli, 179
  fast and slow track to, 188–89
  functions of, 197
  in guiding choices in ambiguous and unpredictable situations, 9
  hyperactivation of, xxix
  in motivational system, 147, 148
  in PTSD, 16
  as relevance detector, 22
  slowing down fast track to, xxvii
  still face impact on, 36
  in threat detection, 176
  tracks to activate, 22–25, 23*f*
Analyze This, 1, 12
anger, 3
anterior cingulate cortex (ACC), 8*f*, 9, 16
  addiction effects on activating, 157
  in conflict resolution, 22
  depression and, 221–22, 232–33
  in detecting social and physical pain, 231–32
  empathy and, 32, 254
  pain and, 32
anterior insula, 7–8, 16
  empathy and, 32
  pain and, 32
antianxiolytic(s)
  exercise as, 126
antibiotics
  dysbiosis related to, 109
antidepressant(s)
  exercise as, xxv, 126
anti-inflammatory cytokines, 96
anxiety, 3. *see also* anxiety disorders
  ACEs impact on, 79
  addiction related to, xxvi, 156
  causes of, 122
  depression and, 228
  exercise in reducing, 129
  focalized, 186–88
  generalized, 181–83 *see also* generalized anxiety
  insecure attachment and, 48
  organized, 175
  stress and, xxvii
  telomere length and, 87–88
  unhealthy lifestyle behaviors and, xxiv
anxiety disorders
  acceptance and commitment therapy for, 266
  amygdala effects of, xxix
  anxiety organized in, 175
  as autostress disorders, 175–89
  emotional chaos self-organizing into, 181
  exposure exercises for, 5
  focalized *vs.* generalized, 181
  formation of, xxvii
  generalized, 181–83 *see also* generalized anxiety disorder (GAD)
anxiety sensitivity, 129
  benzodiazepines and, 186–87
anxiolytic(s)
  aerobic exercise as, xxv
anxious memories, 229–30
apoptosis
  defined, 172
appendix, 107
appetite
  cortisol and, 171
  sleep loss effects on, 133
appetite cycle
  reward system and, 125
approach
  avoidance vs, 234–37
Arden, J.B., ix–xi, xix, 82–84
arginine vasopressin, 80, 166
Aristotle, 139, 152
"as if"
  acting, 189
astrocyte(s)
  in immune system, 98
  PICs released from, 98
astrology
  genetic, 66
atherosclerosis
  increase in, 126

INDEX

atrial natriuretic peptide, 129
attachment
  adult, 47–49
  avoidant patterns of, 56
  disorganized *vs.* secure, 44–49
  endogenous opioids in, 43
  implicit memory of, 41
  insecure, 48
  mental operating networks and, 49–51
  neurochemistry of, 42–43
  overlapping terms for, 42
  oxytocin in, 43
  in predicting adult capacity for intimacy and conflict resolution, 48
  secure *see* secure attachment
  self-soothing and, 79
  variations in, 47–49
attachment research
  interpersonal neurobiology and, xix
attachment styles
  assessment of, 48
  in coping with stress, 42
  ethnic, 55–56
  social self and, 41–44
attention
  contemplative, 256–67 *see also* contemplative attention
  excessive, 39
  reorienting of, 214
  shift in, 214
attentional disorders
  underdevelopment and underactivity in executive network and, 10
attentional flexibility
  meditation in improving, 258
attention deficit disorder
  case example, 51
attention synchronization, 264
attitude(s)
  case example, 249–53
  emotional outcomes related to, 250
  positive, 250–51
  positive psychology and, 249–50
attractor
  defined, 246
autoimmune disorders
  ACE study scores impacting likelihood of, 65
  as bidirectional causal interactions with psychological disorders, 102–3
  chronic inflammation and, 97
  development of, 95
  feedforward loops and, 199
  prevalence of, xxiv
  psychological disorders related to, 94, 102–3
  types of, 97
  unhealthy lifestyle behaviors and, xxiv
automatic thoughts, 189
autonomic-inflammatory reflex
  stress-induced, 198
autostress, 163–89. *see also* stress
  case example, 163–89
  stress and, 163–89
autostress disorders, xxvi–xvii, 175–89. *see also specific disorders, e.g.*, generalized anxiety disorder (GAD)
  anxiety disorders as, 175–89
  balancing executive and salience networks in, 188–89
  case example, 163–89
  development of, 176
  exposure *vs.* avoidance in, 177–80
  focalized anxiety, 186–88
  GAD, 181–83
  generalized anxiety, 181–83 *see also* generalized anxiety; generalized anxiety disorder (GAD)
  sensitization/priming for, 176–77
  slow and fast track related to, 188–89
  stress response systems effects of, xxvii
  stress transitioning into, 177
  top-down, bottom-up, and side-to-side integration in, 180–81
  worry as cognitive avoidance in, 183–85
avoidance
approach *vs.*, 234–37
  cognitive, 183–85
  exposure *vs.*, 177–80
  in GAD, 183
  hypervigilance *vs.*, 205, 205*f*
  trauma-related, 209–14
avoidant behaviors
  depression and, 234–37
avoidant children
  as dismissive adults, 48–49

avoidant patterns of attachment, 56
awakening
  mid-sleep cycle, 136–37
awareness
  emotional, 4
  meta-, 257

bacterial pathogens
  norepinephrine in growth of, 113
bacterium
  in gut lining, 111
Bacteroidetes, xxv, 108
"bad genes"
  case example, 60–61
bad habits
  case example, 157–58
  changing, 157–58
bad health
  psychological stressors related to, 163–64
balance
  energy, 100, 101
basal ganglia
  PICs effects on, 225–26
BDNF. *see* brain-derived neurotrophic factor (BDNF)
behavior(s)
  avoidant, 234–37
  in changing biology, 70
  defensive, 193–94
  emotional, 43
  gambling, 142–45, 144*f*
  health-related *see* behavioral health
  impact on genetic processes, xxiii
  impulsive, 141
  lifestyle, xxiii, xxiv
  passive-aggressive, 39
  positive, 250–51
  prosocial, 43
  telomere length impact on, xxiii
  withdrawal, 234–37
behavioral activation technique
  within cognitive behavioral therapy, 236
behavioral health, 64–66, 92–93
  case example, 91–92
  CDC on, 92–93
  continuum of, 149–50
  physical and mental health comprising, 93
  psychotherapy as, 92
behavior–gene interactions, xxii–xxiv, 59–90
  behavioral health and, 64–66
  case examples, 59–90
  demise of genetic astrology, 66–89
  developmental genomics in, 78–82, 81*f*
  epigenetics in, 69–73
  health consequences of ACEs, 61–64
  in identical twins, 59–90
  introduction, 59–61
  "no blank slate," 73–76
  resilience and, 84–87 *see also* resilience
  substance abuse and, 76–78
  telomeres, health, and longevity, 87–89
  transgenerational trauma and, 82–84
being present in the moment, xxi
belief(s)
  in alleviating suffering, 244
  core, 189
  in "higher power," 257
  in interdependence, 257–58
  mental health and, 248
  placebo–nocebo, 246–48
  placebos and, 244–53
  power of, 244–53
  psychotherapy-related, 247–48
belly fat
  dementia related to, 101
benzodiazepine(s)
  anxiety sensitivity associated with, 186–87
  depression related to, 129
beta-amyloid, 131–32
beta-endorphin(s)
  functions of, 166
  release of, 179
Bettelheim, B., 56
Bifidobacterium infantis, 124
bilateral-reprocessing therapies, 212
biofeedback instruments
  in attaining deeper states of relaxation, 264
biological factors
  placebo effects interacting with, 248
biological feedback loops
  environmental impact on, 71
biology
  factors in changing, 70

INDEX

"blank slate," 73–76
blood glucose levels
  case example, 91–92
blood pressure
  exercise effects on, 129
blood sugar drop
  symptoms following, 117
BMI. *see* body mass index (BMI)
Bodhisattva peace training, 261–62
bodily feelings
  as emotions, 4
body(ies)
  carbohydrates effects on, 117
  emotion and, 3–4
body mass index (BMI)
  Akkermansia level and, 111
  in obesity, 100–1
body sensations
  panic disorder and, 5
  therapy emphasizing, 211–14
body temperature
  regulation of, 136–37
borderline personality disorder
  dialectical behavior therapy for, 266
"boring the worry circuit," 184
bottom-up patterns, 3
  in autostress disorders, 180–81
  emotions-related, 4–5
  kindling motivation through, 159–61
  in mind–brain–gene interactions, 112
  right/left balancing with, 197–98
  unbalances in, 176
bottom-up sensations, xxix
  emotions constructed from, 7–8, 8*f*
Bowlby, J., 41
brain
  activation of inflammatory pathways in, 98–100, 99f
  CEO of see executive network
  chronic illnesses effects on, 155–56
  emotionality as intrinsic function of, 4
  exercise in enhancement and repair of, 130–31
  gaps between mind and, xix
  gene expression in, 73
  gene expression pathways in, 76
  generosity-associated areas of, 253
  genes expressed in, 67
  gut and, 105–6, 111
  in helping essential fatty acids facilitate second-messenger system, 123
  immune system interactions with, 94, 98–100, 99f
  inhibiting neurotransmitters in, 76
  opioid receptor activity in, 79
  placebo effects in, xxix, 244–53 see also placebo effect(s)
  pleasure centers of, 9
  "second," 105–6
  sleep in maintenance of, 132
  social, xxii, 30–31
  "story" see default-mode network (DMN)
  underactive areas of, 242
brain–body feedback loops, 7–8, 8*f*
brain-derived neurotrophic factor (BDNF), 43, 75
  in brain enhancement and repair, 130
  as "brain fertilizer," 84
  depression and, 220
  in energy balance, 101
  functions of, 122, 220
  LPS effects on, 110–11
  in neural plasticity, neurogenesis, memory, energy balance, and mood, 100
  obesity effects on, 101
  omega-3s and, 122
  simple carbohydrates impact on, 119
brain-derived neurotrophic factor (BDNF) gene expression, 84
"brain fertilizer"
  BDNF as, 84
brain functions
  omega-3 essential fatty acids in, 121–22
brain growth
  insulin-like growth-factor and, 85
brain health
  exercise and, 130–31
  factors enhancing, 88
brain maturation
  insulin-like growth-factor and, 85
brain networks. *see* social brain networks; *specific types, e.g.*, default-mode network (DMN)
"brand names"
  transcending, xxviii
breathing
  sleep-disordered, 135
"brown-out," 100

calming effect
  oxytocin and, 43–44
Canadian Community Health Survey, 65
cancer(s)
  diet high in saturated fat and, 120
  increase in, 126
  WHO on, 120
cannabinoid circuitry
  in enjoying, 142
cannabis
  case example, 76–77
  psychotic symptoms after smoking, 76–77
cannabis smoking
  alcohol use *vs.*, 77–78
Cannon, W., 4
carbohydrate(s)
  BDNF effects of, 119
  bodily effects of, 117
care, 3
caregiver(s)
  telomere length in, 87
catecholamine(s), 173
catechol-O-methyltransferase (COMT)
    protein, 77
causation
  vulnerability *vs.*, 74
CDC. *see* Centers for Disease Control and
    Prevention (CDC)
cecum, 107
cell(s). *see also specific types, e.g.*, spindle
    cells
  dendritic, 106
  dopamine-containing, 125
  fat, 100–2
  of gut, 106, 107
  immune, 107
  impact on genes, 71
  natural killer, 168
  responding to internal and external
    conditions, 71
"cells that fire out of sync loss their link,"
    207
Centers for Disease Control and Prevention
    (CDC)
  on childhood adversity effects on adult
    health, 61–62
  on health behaviors, 92–93
  on obesity, 100–1
  on suicide, 224
  on usefulness of ACE questions in health
    survey, 64
central adiposity
  dementia related to, 101
central executive. *see* executive network
central nervous system (CNS)
  chronic inflammation effects on, 225–26
CEO of brain. *see* executive network
CERs. *see* conditioned emotional responses
    (CERs)
Chagdud Tulku Rinpoche, H.E., 261
change(s)
  in gene expression, 71
  genetic, 69–73
  intention as predictor of actual, 159–60
  lifestyle, 105
  motivation to, 160
  psychotherapy-related, x–xi
change talk, 160–61
chaos
  emotional *see* emotional chaos
  mind–body, 164–75
Charcot, J-M, 61, 211
check-and-balance system, 197
"checked out," 209
chemical dependency
  DSM on, xxvi
childhood maltreatment
  elevated levels of inflammation related
    to, 65
childhood psychological stress
  adult health problems related to, 61–62, 64
"child of a rageaholic," 189
child–parent dyad
  as self-organizing system, 41
  stories in, 53
Child Protective Services
  on leaving child unattended, 56
children
  avoidant, 48–49
  co-constructing narratives with, 54–55
  excessive attention and coddling effects
    on, 39
  inflammation in, 97
  leaving unattended, 56
  maternal separation effects on, 45
  narcissistic, 39
  secure attachment in, 47
  spoiled, 39

choice(s)
  amygdala in guiding, 9
cholesterol, 172
cholesterol levels
  case example, 91–92
Christian contemplation, 261
Christian Meditation, 256
chromatin, 72
chromosome(s), 72
chronic illnesses
  brain changes related to, 155–56
  immune system effects of, 94
  obesity and, 100
  physical inactivity and, 126
chronic inflammation, 96–97. *see also* inflammation
  acute inflammation vs., 102
  adverse effects of, 98, 102
  anxiety, depression, and cognitive defects related to, xxiv
  autoimmune disorders and, 97
  causes of, 223–24
  CNS effects of, 225–26
  dementia related to, xxiv
  health and mental health problems related to, xxiv
  leaky gut and, 118
  neurodegenerative diseases and, 97
  PICs and, 96–97
  short-term inflammation vs., 96
  sickness behavior related to, 224–27
chronic insomnia
  shorter telomeres and, 132
chronic pain
  MBSR for, 266
chronic stress
  case example, 103–4
  depression and, 199
  hypoactivation of HPA axis related to, 173
  immune system effects of, 94
  impact of, 85
  pathogens' growth related to, 113
  poor health related to, 174–75
  telomere length and, 87–88
circadian rhythm
  for cortisol, 136
  healthy, 136–37
  resetting, 135–37
closed systems
  open systems *vs.*, xxix
close relationships
  longevity related to, xxii
CNS. *see* central nervous system (CNS)
cocaine
  as "rock," 154
co-constructive narratives, 54–55
coddling
  excessive, 39
cognition(s)
  chronic inflammation effects on, xxiv
  sleep impact on, xxv
  top-down, 7–8, 8*f*
cognitive avoidance
  worry as, 183–85
cognitive behavioral therapy, 181
  behavioral activation technique within, 236
  to metacognitive models, 185
  for obsessive-compulsive disorder, 266
  variant of, 184
cognitive deficits
  causes of, 99
cognitive development
  maternal depression impact on, 37–38
cognitive distortions
  types of, 181
cognitive reappraisal, 160
"comfort cocoon," 139
comfort foods, 124
communication patterns
  implicit, 51–52
compassion
  contemplative, 260–62
  cultivation of, 257–58
  empathy and, 254
  expressing, 254
  importance of, 260
  mental health effects of, xxx
  self-, 254–55, 262
compassionate engagement, 261
competence
  cultural, 57
  social, 33
complex adaptive systems
  feedback loops in, 268
complex decision making, xxi

complexity theory, xix–xxi
complex trauma
  adverse effects of, 198
computer games
  mental health effects of, xxvi
COMT gene
  psychoses related to, 77
COMT protein. *see* catechol-O-methyl-transferase (COMT) protein
conditioned emotional responses (CERs), 208
  neutralizing, 208
  traumatic, 208–9
conflict
  internal, 232–33
conflict resolution
  ACC in, 22
  attachment in predicting adult capacity for, 48
conscious emotion
  fear as, 176
conscious reasoning
  in PFC, 145
contemplation, 256, 262
  Christian, 261
  effect of, 261
contemplative attention, 256–67
  mental operating systems and, 258–60, 259f
contemplative compassion, 260–62
contemplative meditation
  calming factors in, 257–58
  history of, 256–57
  self-compassion in, 262
contemplative prayer, 256
continuous positive airway pressure (CPAP) device
  for sleep apnea, 135
conversion disorder, 61
coping
  stress-related, 42
core beliefs, 189
"core sleep," 132
cortex
  anterior cingulate *see* anterior cingulate cortex (ACC)
  in emotional awareness, 4
  in exposure to stress-provoking stimuli, 180
  fear and, 5
  posterior parietal, 11, 11*f*
  prefrontal *see* prefrontal cortex (PFC)
corticotropin-releasing hormone (CRH), 80, 166, 167, 173
  in agitated depression, xxix
  hypersecretion of, 228
cortisol, 168, 170–74
  ACEs effects on, 141
  amygdala effects of excessive, 174
  appetite and, 171
  circadian rhythm for, 136
  functions of, 170–71
  high levels of, 130
  impact on hippocampus and frontal lobes, xxv
  type 2 diabetes related to, 174–75
cortisol receptors
  factors impacting gene regulation of, xxiii
counterchange talk, 160
Cozolino, L., xi, xix
CPAP device. *see* continuous positive airway pressure (CPAP) device
crack houses, 154
C-reactive protein (CRP) levels
  case example, 91–92
  inflammation related to, 118
  measurement of, 97
  sleep dysregulation effects on, 135
CRH. *see* corticotropin-releasing hormone (CRH)
critical incidence debriefing
  adverse effects of, 195
cross-sensitization
  in autostress disorders development, 176–77
CRP. *see* C-reactive protein (CRP)
Crystal, B., 1
cue(s)
  in alcoholism, 154
  emotions as, 208–9
  in flashbacks, 204
  gambling behaviors and, 142–45, 144*f*
cultural competence
  in psychotherapy training, 57
culture
  impact on social self, 55–57
culture-specific "display rules"
  emotions-related, 3

INDEX

cytokine(s)
  anti-inflammatory, 96
  described, 96
  pro-inflammatory *see* pro-inflammatory cytokines (PICs)

Darwin, C., 3, 27, 68, 69
  theory of natural selection of, 68–69
Dawkins, R., 68
daydreaming, xxi
  non-sleep-time, 12
D2 dopamine receptor
  A1 allele of, 140–41
death march, 82–83
debriefing
  critical incidence, 195
de Chardin, P.T., 269
decision making
  complex, xxi
  mental operating networks in, 21
declarative memory, 17
  recalled through executive network and DMN, 20
declarative memory system
  as explicit memory system, 17
"deer-in-the-headlights" mode, 194, 209
default-mode network (DMN), xxi, 6, 12–15, 13*f*
  abnormal "resting state" activity associated with, 133
  ACEs impact on, 14
  case example, 50
  depression and, 14–15
  described, 13–15
  development of, 14, 50
  diet effects on, 117
  exercise in decreasing and normalizing, 127
  explicit memories recalled through, 20
  fantasies generated in, 14–15
  function of, 14–16
  hyperconnectivity of, 237–38
  long-term memory systems in, 21
  in meditation, 258–60, 259f
  PICs effects on, 227
  in PTSD, 16
  salience network with, 50, 238–39
  sleep deprivation effects on connectivity of, 133
defense(s)
stages of, 193
defensive behaviors
  traumatic events and, 193–94
dementia(s)
  central adiposity and, 101
  chronic inflammation and, xxiv
  exercise in lowering risk of, 124
  obesity and, 101
demyelination
  causes of, 221
dendritic cells, 106
deoxyribonucleic acid (DNA)
  described, 71
depression, 217–42
  abnormal DMN activity related to, 127
  ACC and, 221–22, 232–33
  ACEs impact on, 79, 219–20
  ACE study score and, 79
  acute stress and, 199
  addiction related to, 156
  amygdala activity and, 229–30
  amygdala effects of, xxix
  anxiety and, 228
  approach *vs.* avoidance related to, 234–37
  avoidant behaviors and, 234–37
  BDNF and, 220
  as being stuck in psychological rigidity, xxix
  benzodiazepines and, 129
  biological links with inflammation and ACEs, 223
  broad approach to, 241–42
  causes of, 122, 222
  chronic stress and, 199
  cognitive and affect regulatory deficits with, 232–33
  conditions related to, xxviii
  described, 227–28
  DLPFC and, 221–22
  DMN and, 14–15
  dopamine and, 228–29
  effort to reestablish pleasurable activities in management of, 236
  epigenetics and, 221
  exacerbation of, 222
  exercise in lowering risk of, 124, 127
  between extremes, 239–41

in families, 223
fibroblast growth factor and, 220
GDNF and, 220
gene–environment interactions in, 219–20
genetics and, 221
hostility and, 224
5-HTTLRR in vulnerability to, 75–76
hypocorticolism and, 173–74
illnesses related to, xxviii
in infants, 36–38
inflammation and, 222–24
LPS and, 110
maternal *see* maternal depression
mindfulness-based cognitive behavioral therapy for, 266
neurotrophic factors and, 220–22
NGF and, 220
parental, 35–38
PICs effects on, 226–27
postpartum, 35–38
prevalence of, 218
pseudo-, 10
revolving internal conflict and, 232–33
risk factors for, 222–23
sickness behavior related to, 224–27 *see also* sickness behavior
social stress and, 230–32
stress and, 219, 228–32
stress–depression synergy, 227–30, 229*f*
suicide related to, 224
telomere length and, xxiii, 87–88
transcending rigidity in management of, 217–42
treatment of, 240–42
type 2 diabetes and, 102, 224
types of, 218–37
unhealthy lifestyle behaviors and, xxiv
withdrawal behaviors and, 234–37
in women *vs.* men, 218
"depression gene," 74–75
depression switch, 233
deprivation
CRH levels and, 228
depression related to, 223
early, xxii
maternal, 228
pathophysiological syndromes resulting from, 112–13
Desert Fathers, 256
Deshimaru, T., 91, 113
destructive habits
ACEs and, 63–64
detachment
in dissociation, 209–10
extreme, 209–10
levels of, 209–10
mild, 209–10
moderate, 209–10
trauma-related, 209–14
development
cognitive, 37–38
of DMN, 50
healthy, 39
left hemisphere during, 28–29, 29*f*
right hemisphere during, 28–29, 29*f*
developmental genomics, 78–82, 81*f*
diabetes mellitus
memory effects of, 119
diabetes mellitus type 2. *see* type 2 diabetes
"diabetes type 3"
Alzheimer's disease as, 102
Diagnostic and Statistical Manual of Mental Disorders (DSM), 66
on chemical dependency, xxvi
dialectical behavior therapy, xxx, 265–66
for borderline personality disorder, 266
diet
of ancestors, 116–17
DMN effects of, 117
executive network effects of, 117
high in saturated fat, 120
immune system effects of, 105
inflammation and, 124–25
mainstream, 116–25
Mediterranean, 123–24
mental health effects of, xxv
Okinawan, 124, 125
Paleolithic, 117
perils of mainstream, 116–25
"plants plus," 124
in self-maintenance, 115–25
stress tolerance related to, 116, 124
Western, 116–25
WHO on, 120
direct pathway
in response to dopamine, 148–51

discomfort
  intestinal, 105, 107, 109–10
dismissive adults
  avoidant children as, 48–49
dismissive attachment style, 48–49
  case example, 49, 50
disorganized attachment
  dissociative problems related to, 47
dissociation
  detachment in, 209–10
  trauma-related, 209–14
dissociative problems
  disorganized attachment and, 47
distortion(s)
  cognitive, 181
DLPFC. *see* dorsolateral prefrontal cortex (DLPFC)
DMN. *see* default-mode network (DMN)
DNA. *see* deoxyribonucleic acid (DNA)
DNA methylation
  levels of, 79
docosahexaenoic acid
  brain function effects of, 122
dopamine, 123
  ACEs effects on, 141
  in addiction, 142–43
  circuits responding to, 148
  depression and, 228–29
  described, 142
  medium spiny neurons responding to, 148
  motivation and, 142
dopamine activation
  addiction resulting from, 144*f*
dopamine activity
  changes in, 152
dopamine circuits
  in wanting, 142
dopamine-containing cells, 125
dopamine firing
  case examples, 152–54
  intensity and rate of, 152–53
dopamine neurons, 146
  firing of, 142–45, 144*f*
  function of, 152
dopamine receptor(s)
  D2, 140–41
dopamine receptor gene (DRD4), 75
dopamine release, 125
  reward circuit in, 146
dopamine reward system
  enthusiasm and, 253
dorsolateral prefrontal cortex (DLPFC), 10–11, 11*f*
  depression and, 221–22
double helix, 72
DRD4. *see* dopamine receptor gene (DRD4)
drug(s)
  addiction related to, 154–55
  methods of taking, 154–55
  repeated use of, 155
DSM. *see* *Diagnostic and Statistical Manual of Mental Disorders* (DSM)
Dunedin Multidisciplinary Health and Development Study, 65, 74, 77
durability
  resiliency as, 170
Dutch Hunger Winter, 69–70
dysbiosis, 109–11, 125
  antibiotics use and, 109

early deprivation
  impact on affect stability, xxii
early-life trauma
  depression related to, 223
early nurturance
  stress tolerance effects of, 59
early psychological adversity
  physical health problems related to, 61
early stress
  effects of, 112–13
  pathophysiological syndromes resulting from, 112–13
effort-driven reward circuit, 236
eicosapentaenoic acid (EPA)
  brain function effects of, 122
Ekman, P., 3
EMDR. *see* Eye Movement Desensitization Reprocessing (EMDR)
emotion(s)
  arising from limbic system, 3
  bodily feelings as, 3–4
  from bottom-up sensations and top-down cognitions, 4–5, 7–8, 8f
  common set of, 3
  conscious, 176

construction of, 4
as cues, 208–9
culture-specific "display rules" related to, 3
emergence of, xxi, 7–9, 8f
expressions of, 3–4
feelings as, 4
generation of, 4
intensity of, 200
interaction between explicit and implicit systems influenced by, 200
secondary, 203–4
"self"-organization and, 3–5
thoughts and, 3–4
top-down interactions, 4–5
emotional awareness
cortex in, 4
emotional behaviors
oxytocin as regulator of, 43
emotional blunting
hypervigilance *vs.*, 205, 205*f*
emotional chaos
patterns of, 175
self-organizing into anxiety disorders, 181
stress and, 164–75
"emotional conflict adaptation effect," 21
Emotional Freedom Therapy, xxviii, 212
emotional implicit memory, xxvi
as immediate, 22
emotionality
as intrinsic function of brain, 4
emotional memory
sleep deprivation and, 134
as subsystem of implicit memory system, 18–20
emotional neglect
case example, 41–42, 60–61
in "self"-organization, 42
emotional nourishment
in healthy development, 39
emotional nurturance
lack of, 36
emotional outcomes
attitudes and, 250
emotional regulation
implicit forms of, 21
emotional states
positive, 252
emotion/cognition interface
in stress-related learning and memory, 76
empathy
ACC and, 32, 254
action-oriented, 254–55
anterior insula and, 32
compassion from, 254
cultivating, 31–33
endogenous opioids
in attachment, 43
endorphin(s)
exercise effects on, 129
energy
"storing" of, 117
energy balance
BDNF in, 100, 101
energy expenditure
obesity impact on dysregulation of, 101
energy intake
obesity impact on dysregulation of, 101
engagement
compassionate, 261
enhancer(s), 67
enjoying
opioid and cannabinoid circuitry related to, 142
"enough is never enough," 251
ENS. *see* enteric nervous system (ENS)
enteric nervous system (ENS), 105–13, 124
described, 106
enthusiasm
dopamine reward system and, 253
environment
impact on biological and psychological feedback loops, 71
environment–gene interactions, 72
psychotherapy in influencing, 59
EPA. *see* eicosapentaenoic acid (EPA)
epigenesis
described, 70
epigenetic effects
occurrences of, 73
social factors related to, 78
epigenetics, 69–73
for better or worse, 75
defined, 70
depression and, 221
described, xxiii

epigenetics (*continued*)
developmental genomics impact on, 78–82, 81f
on effects of quality of care early in life, 59
in enhancing effects of nurturance or exacerbating effects of ACEs, 75
of exercise, 128
health problems related to, 59
in mind–brain–gene feedback loops, 58f
neurohormonal effects of, 86
in revolutionizing biological science, health care, and psychotherapy, 70, 71
substance abuse and, 76–78
epinephrine, 168
release of, 165
episodic memory(ies), 20–21
secondary emotions associated with, 203–4
Ericksonian hypnosis, 247
essential fat(s), 120–22
essential fatty acids, 119–20
benefits of, 122–23
in helping brain facilitate second-messenger system, 123
omega-3 *see* omega-3 essential fatty acids
essential neurochemical cornucopia, 123–24
estrogen
epigenetic changes to, 86
ethnic attachment styles, 55–56
"everything but the kitchen sink gene," 74
evolution
in understanding mind–brain–gene feedback loops, 68
evolutionary theory
gene-related, 68
excessive fat cells
mental health consequences of, 100–2
excitement
"rush" of, 143–45
executive network, xxi, 9–12, 11*f*, 100
central, 6
diet effects on, 117
explicit memories recalled through, 20
function of, 15–16
long-term memory systems in, 21
in meditation, 258–60, 259*f*
salience network integration with, 146–47, 188–89
underdevelopment and underactivity in, 10
exercise(s), 125–31
aerobic, xxv
as antidepressant, xxv, 126
anxiety reduction related to, 126, 129
benefits of, xxv, 126, 127
blood pressure effects of, 129
brain health related to, 130–31
for depression, 127
endorphins effects of, 129
epigenetics of, 128
exposure, 5
GABA effects of, 129
gene activation after, 128
immune system effects of, 105
in lowering risk of dementias, 124
mental health effects of, 126
neurohormones effects of, 129
neuropeptides effects of, 129
neurotransmitters effects of, 129
in normalizing gene expression, 128
in pain modulation, 166
relaxation, 182–83, 257–58
serotonin levels effects of, 129
somatic, 182–83
studies of, 127
expectancy, 3
"expectancy set," 247
expectation(s)
managing of, 152–53
moderating, 152–53
outcome of health care treatments related to, 244
explicit memory(ies)
durability of, 19
hippocampus in, 21
mental networks and, 18*f*
modified over time, 200
procedural memory providing clues for, 20
recalled through executive network and DMN, 20
semantic, 20–21
explicit memory system, 199
declarative memory system as, 17
implicit memory system interacting with, 200
implicit memory system *vs.*, 18, 22

INDEX

exposure
  avoidance *vs.*, 177–80
exposure exercises
  for anxiety disorders, 5
expression
  gene *see* gene expression
extreme detachment, 209–10
extrinsic motivation
  secondary reinforcers in, 161
Eye Movement Desensitization Reprocessing (EMDR), xxviii, 212

face-to-face play, 56–57
Facial Action Coding System (FACS), 3
FACS. *see* Facial Action Coding System (FACS)
failure to thrive
  lack of emotional nurturance and, 36
family(ies)
  depression in, 223
famine
  genetic effects of, 70
fantasy, xxi
  DMN and, 14–15
fast track
  to amygdala, 22–25, 23*f*
  slowing down, 188–89
fat
  belly, 101
fat(s)
  essential, 120–22
  saturated, 120
fat cells
  excessive, 100–2
fatty acids
  essential, 119–20
  short-chain, 125
  trans-, 120–21
fear, 3
  as CER, 208
  as conscious emotion, 176
  cortex and, 5
  panic disorder and, 5
feedback loop(s)
  brain–body, 7–8, 8f
  causes of, xx
  in complex adaptive systems, 268
  described, xx
  functions of, xx
  between homeostatic system and stress systems, 164
  of immune system, 94–95, 95f
  within mind–brain–gene continuum, xviii, 269
  negative, xx
  positive, xx
  in promoting growth and maintaining homeostasis, xx
  psychological, 71
  underlying stress and immune responses to threat, 167f
feedforward loop(s)
  autoimmune disorders and, 199
  trauma symptoms–related, 196–97
feeling(s)
  in awareness of time and self, 267
  bodily, 4
  gut, 32, 105–13
  implicit, 53
  intrusive trauma-related, 199
  intuitive, 32
  "nervous gut," 107
  salience network in promoting, 32
"feeling network." *see* salience network
Felitti, V., 61, 64
female hysteria, 61
fibroblast growth factor
  depression and, 220
fight-or-flight response, 165, 166, 193
  norepinephrine related to, 168
Finley, J., 256–57
Firmicutes, xxv, 108, 110, 118
flashback(s)
  CERs triggering, 209
  cued by implicit memories, 204
  emotional intensity associated with, 201
focalized anxiety, 186–88
  case example, 186–88
  interoceptive exposure techniques for, 187–88
food(s)
  comfort, 124
forgiveness
  mental health effects of, xxx
fragmented sleep
  trauma and, 134

frame of reference
side-to-side, 177–78
Framingham Heart Study, 65
Francis of Assisi, 217
Frank, J., 247
Frankl, V., 215
Freud, S., 61
frontal lobes
cortisol impact on, xxv
fructose
adverse effects of, 118
fruit drinks
in Western diet, 119

GABA. *see* gamma-aminobutyric acid (GABA)
GAD. *see* generalized anxiety disorder (GAD)
gambling
addiction and, 142–45, 144f
behaviors related to, 142–45, 144f
mental health effects of, xxvi
gamma-aminobutyric acid (GABA), 123
exercise effects of, 129
functions of, 76
gamma waves, 264
gastrointestinal system
as "second brain," 105–6
GDNF. *see* glial cell–derived neurotrophic factor (GDNF)
GDNF gene. *see* glial cell-derived neurotrophic factor (GDNF) gene
gene(s). *see also specific types*
ACEs related to, 74
activation after exercise, 128
"bad," 60–61
cells impact on, 71
composition of, 71–72
COMT, 77
defined, 67
"depression," 74–75, 221
evolutionary theory involving, 68
factors impacting, xxiii
famine effects on, 70
GDNF, 85
how we think about, 60–61
in human genome, 67
interacting with environment, 72
junk, 67
long-term modification of, 70
"metabolic/mitochondrial enzyme," 128
"metabolic-priority," 128
NGF, 85–86
protein-coding, 67
serotonin-transporter *see* serotonin-transporter gene (5-HTTLRR)
"stress-response," 128
stress tolerance effects of, xxiii
"vulnerability," 75
gene–behavior interactions, 59–90. *see also* behavior–gene interactions
gene–environment interactions
depression related to, 219–20
gene expression or suppression related to, xxiii
psychotherapy in influencing, 59
gene expression, 67
BDNF, 84
in brain, 73
changes in, 71
daily exercise in normalizing, 128
described, 69
developmental factors in, 78
gene–environment interactions and, xxiii
methylation in, xxiii
psychological experiences effects on, 61
psychotherapy effects on, xix, 59, 78
self-maintenance system in, 115
social factors and, 78, 82
types of, 72–73
gene functioning
stressful events impact on, 76
generalized anxiety, 181–83. *see also* generalized anxiety disorder (GAD)
case example, 181
formation of, xxvii
generalized anxiety disorder (GAD), 181–83. *see also* generalized anxiety
avoidance associated with, 183
described, 181–82
5-HTTLRR in vulnerability to, 75–76
hypervigilance in, 183–84
intolerance to ambiguity in, 183–84
relaxation methods for, 182–83
somatic exercises for, 182–83

generosity
  brain areas associated with, 253
gene suppression, 59, 67
  acetylation in, xxiii
  gene–environment interactions and, xxiii
  mode of, 72–73
genetic astrology
  demise of, 66–89
  described, 66
genetic change, 69–73
  rapid, 70
genetic disorders
  causes of, 72
genetic effects
  complexity of, 67
genetic processes
  behaviors impact on, xxiii
genetic reductionism
  embracing of, 67–68
  limitations of, 66–67
  to mind–brain–gene feedback loops, 89
genome
  defined, 71
genomics
  developmental, 78–82, 81f
  social, 78
ghrelin, 133
"giving is receiving," 252
GL. *see* glycemic load (GL)
glial cell(s), 98, 131
glial cell-derived neurotrophic factor (GDNF)
  decreased methylation related to, 85
glial cell–derived neurotrophic factor (GDNF)
  in brain enhancement and repair, 130
  depression and, 220
  functions of, 220–21
global happiness reports, 250
glucose levels
  balancing of, 118
  GL in measuring, 119–20
glutamine, 123
glycemic load (GL), 119–20
  glucose levels measured by, 119–20
glycogen, 117
glymphatic system, 131
"good-enough" parenting, 38–41
  benefits of, 39
  case example, 38–39
"go somewhere else," 209
gratification
  deferring of immediate, 147
gratitude
  benefits of, 251–53
  mental health benefits of, xxx, 252–53
  positive emotional states and, 252
  practicing, 250
greater gain
  medium spiny neurons and, 147–52
grief/panic, 3
group child-rearing practices
  on kibbutz, 56
growth
  feedback loops in promoting, xx
growth-factors
  insulin-like, 85
gut
  immune cells of, 107
  immune system of, 106–7
  leaky, 109–10, 118
  microorganisms in, 107–9
  PICs effects on, 107, 110
  as "second brain," 105–6
  site of, 106
  size of, 106
gut–brain interactions, 106, 111, 124
gut feelings, 105–13
  described, 111
  salience network in promoting, 32
"gut feelings"
  nervous, 107
gut hormones, 124–25
gut lining
  bacterium relationship to, 111
  supporting of, 111
gut microbes
  in norepinephrine modification, 113
gut microbiota, 108–9

habit(s), 139–61
  bad, 157–58
  becoming addictions, xxvi
  case example, 139–40
  changing, 157–58

habit(s) (*continued*)
destructive, 63–64
establishing new, 19–20
forming, 158–59, 158*f*
in mind–brain–gene feedback loops, 138*f*
modifying, 158–59, 158*f*
motivation effects of, xxv–xxvi
neuroscience underlying, xxvi
PFC in forming and modifying, 156–59, 158*f*
procedural memory and, 19
habit circuits
changing, 157–59, 158*f*
happiness
paradox of, 250
"stumbling on," 250
happiness reports
global, 250
health
ACEs impact on, 61–64
ACE study ramifications on, xxii–xxiii
bad, 163–64
behavioral *see* behavioral health
body–mind and, 91–114
brain, 130–31
chronic inflammation effects on, xxiv
factors in addressing overall, xviii
mental *see* mental health
microbiota and, 108
poor *see* poor health
positive relationships impact on, xxii
self-care practices in undermining, xxiv–xxv
health behaviors. *see* behavioral health
health care
decompartmentalized, 64–65
epigenetics in revolutionizing, 70, 71
integrated approach to, 64
psychotherapy interdependence with, xviii–xix
health problems
childhood psychological stress and, 64
early psychological adversity and, 61
epigenetic effects and, 59
health surveys
ACE–related questions in, 64
healthy development
emotional nourishment critical for, 39
"higher power"
belief in, 257
high-sensitivity reparations, 40
"highway hypnosis," 12
hippocampal-frontal memory system, 200
hippocampus, 146, 174
in comparing knowledge, 22
cortisol impact on, xxv
in explicit memory, 21
in exposure to stress-provoking stimuli, 180
GABA effects on, 76
in memory, 130
PICs effects on, 226–27
histone(s), 72
"hitting the wall," 179
Holocaust
offspring of mothers who suffered PTSD associated with, 84
homeostasis
described, 164
factors affecting, 164
feedback loops in maintaining, xx
homeostatic system
feedback loops between stress systems and, 164
hormone(s). *see specific types*
horror
as CER, 208
hostility
depression related to, 224
hot spots
described, 201
HPA axis. *see* hypothalamic-pituitary-adrenal (HPA) axis
5-HTTLRR. *see* serotonin-transporter gene (5-HTTLRR)
Human Genome Project, 67, 108
hyperarousal zone, 205*f*
hypervigilance
avoidance and emotional blunting *vs.*, 205, 205*f*
GAD and, 183–84
hypnosis
calming factors in, 257–58
Ericksonian, 247
"highway," 12
hypoarousal zone, 205*f*

hypocapnic alkalosis, 187
hypocorticolism, 173–74
hypothalamic-pituitary-adrenal (HPA) axis, 80, 130, 166, 170, 172
  disruptions in, 228
  hypoactivation of, 173
hypothalamus
  PICs effects on, 226–27
hysteria
  female, 61

identical twins
  behavior–gene interactions in, 59–90
ill health. *see* poor health
illness(es)
  dysregulating spiral with, 104–5
  physical, 93
  PICs effects of, 105
  telomere length and, 88
immediate gratification
  deferring of, 147
immobilization, 193
immune cells
  of gut, 107
immune responses, 105–13
  to acute stressors, 168
  stress responses in modifying, 164
  to threat, 167*f*
immune system, 93–105
  astrocytes in, 98
  of brain, 98–100, 99*f*
  brain communication with, 94
  chronic conditions effects on, 94
  components of, 94, 95*f*
  damaging effects on, 94
  dysregulation of, xxiv, 93
  feedback loops of, 94–95, 95*f*
  functioning of, 115
  functions of, 94
  of gut, 106–7
  interactions with brain, 98–100, 99*f*
  lifestyle changes impact on, 105
  MBSR effects on, 265
  parasympathetic nervous system relationship with, 165
  sleep loss and disturbances effects on, 134
  stress disrupting, 103, 166–68, 167*f*
implicit communication patterns, 51–52
implicit memory(ies)
  of attachment, 41
  durability of, 19
  emotional, xxvi, 18–22
  factors associated with, 20
  flashbacks cued by, 204
  forming fabric of "self"-organization, 20
  mental networks and, 18f
  nondeclarative, 21
  recalled through salience network, 20
  resistant to change and passage of time, 200
implicit memory system, 18–20, 18*f*, 199–200
  explicit memory system interacting with, 200
  explicit memory system *vs.*, 18, 22
  motivation and, 141
  nondeclarative memory as, 17
  procedural and emotional subsystems of, 18–20
implicit-nonconscious procedural memory system, 211
impulse control problems
  prefrontal deficits related to, 147
impulsive behaviors
  A1 allele of D2 dopamine receptor and, 141
inactivity
  chronic illnesses related to, 126
indirect pathway
  in response to dopamine, 148–51
individuality
  evolving experiences of, 2
infant(s)
  of depressed mothers, 37
  depression in, 36–38
  parental depression effects on, 35–38
infant–mother dyad, 41
infant–parent dyad
  quality of reciprocity within, 39
Infant Strange Situation paradigm, 40, 48, 55–56
infection(s). *see specific types*
inflammation. *see also* chronic inflammation
  acute vs. chronic, 102
  aging effects on, 126
  BDNF effects of, 122
  causes of, xxiv
  in children, 65, 97

chronic see chronic inflammation
conditions impacting, 102
CRP elevation and, 118
depression and, 222–24
described, 96
diet and, 124–25
obesity and, 100
PICs and, xxiv
psychological disorders and, 97
short-term vs. chronic, 96
sickness behavior related to, 224–27 see also sickness behavior
sleep dysregulations and, 135
stress and, 103–5
inflammation–stress connection, 103–5
inflammatory bowel disease
PICs and, 107
inflammatory pathways
brain in activation of, 98–100, 99f
"inflammatory reflex," 165
inheritance of acquired characteristics
theory of, 68
inhibiting neurotransmitters
in brain, 76
insecure attachment
anxiety and mood disorders associated with, 48
insomnia
adverse effects of, 134–35
chronic, 132
shorter telomeres and, 132
"sleep maintenance," 136–37
insula, 7–9, 8*f*
anterior see anterior insula
insulin
role of, 174–75
insulin-like growth factor
brain effects of, 85
play and, 85
insulin resistance
reduced physical activity and, 128
integrated psychotherapy
mental health effects of, xxvi
integration
memory, 200–2
intention to change
as predictor of actual change, 159–60
interdependence
belief in, 257–58
interleukin(s), 96
disorders associated with, 102
IL-1, 98
IL-6, 102, 168, 223
internal conflict
revolving, 232–33
interoceptive exposure techniques
for focalized anxiety, 187–88
interpersonal neurobiology
attachment research and, xix
interpersonal relatedness
overlapping concepts of, 52
interpersonal stress
coping with, 42
interviewing
motivational see motivational interviewing
intestinal discomfort
case example, 105, 107, 109–10
intimacy
attachment in predicting adult capacity for, 48
intrinsic motivation
activating, 161
building on, 161
intrusive trauma-related thoughts and feelings, 199
intuitive feelings
salience network in promoting, 32
isoprostanes, 119
Alzheimer's disease related to, 119

James, W., 1, 2, 4
Janet, P., 211, 213
Jesus, 243
joy
social, 3
junk DNA
described, 67
functions of, 67

Kaiser Permanente Medical Centers, 61
mind–body class at, 264–65
Kandel, E., 59
Keller, H., 191
kibbutz
group child-rearing practices on, 56

INDEX

knowledge
  PFC and hippocampus in comparing, 22
Kundalini meditation, 258–59

Lamarck, J-B, 68–69
  theory of inheritance of acquired characteristics of, 68–69
"latent schizophrenia," 77
leaky gut, 109–10
  chronic inflammation related to, 118
  contributors of, 109–10
lean people
  Akkermansia in, 111
learning
  stress-related, 76
left hemisphere
  during development, 28–29, 29*f*
Leghorn, L., 261
leptin, 118, 133
leptin resistance
  obesity and, 100
life satisfaction
  fleeting and unexpected moments of, 250
lifestyle behaviors
  immune system dysregulation due to, xxiv
  telomere length and, xxiii
  unhealthy, xxiv
lifestyle changes
  immune system effects of, 105
life-threatening situations
  responses to, 191
liking
  opioid and cannabinoid circuitry related to, 142
limbic system
  emotions arising from, 3
Lincoln, A., 239
Lincoln, M.T., 239–40
lipopolysaccharide (LPS), 110, 118
  Alzheimer's disease related to, 111
  BDNF effects of, 110–11
  depression related to, 110
listening
  active, 57
Locke, J., 73, 149
longevity
  close relationships and, xxii
  Mediterranean diet and, 124
long-term dysregulations
  PTSD–related, 198–99
  trauma and, 198–99
long-term memory
  mental networks and, 18*f*
  organization of, 17
long-term memory systems, xxi–xxii
  list of, 18
  neuroplasticity for, 17
  in "self"-organization, 17–22, 18*f*
loss
  social self and, 44–47
"lost their sense of purpose," 213
loving-kindness meditation, 262
low-sensitivity reparations, 40
LPS. *see* lipopolysaccharide (LPS)
lust (sexual excitement), 3

macrophage(s), 95–96
Main, M., 48
malondialdehyde, 120
maltreatment
  elevated levels of inflammation due to, 65
Mandela, N., 168, 216, 249
Man's Search for Meaning, 215
marshmallow test, 146–47
matching
  moderate *vs*. perfect, 38
"material me," 6, 7
maternal depression
  case example, 35–36
  effects of, 35–38
  impact on infants, 35–38
maternal deprivation
  CRH levels and, 228
maternal separation
  effects on child, 45
MBSR. *see* mindfulness-based stress reduction (MBSR)
McGill University, 79
Meaney, M., 79
meaning
  development of, 214–16
meaning-making process
  mind as, xx–xxi
"meaning response"
  placebo response as, 247
medical centers
  mind–body medicine in, 264–65

Medieval Christian mystics, 256
meditation
  contemplative, 256–58, 262
  for depression, 240–41
  described, 258
  DMN in, 258–60, 259*f*
  effectiveness of, 265
  executive network in, 258–60, 259*f*
  health effects of, xxx
  in improving attentional flexibility, 258
  Kundalini, 258–59
  loving-kindness, 262
  mental operating networks and, 258–60, 259*f*
  metta, 262
  neural correlates of, 263–64
  salience network in, 258–60, 259*f*
  styles of, 258–59
  Tibetan, 258–59
  Vipassana, 259
  Zen, 259
meditation-augmented therapy, 263–67
Mediterranean diet, 123–24
  longevity related to, 124
  in telomere maintenance, 124
medium spiny neurons
  greater gain related to, 147–52
melatonin, 135
memory(ies)
  anxious, 229–30
  BDNF in, 100
  declarative, 17, 20
  depressed, 229–30
  diabetes effects on, 119
  emotional, 134
  emotion/cognition interface in, 76
  episodic, 20–21, 203–4
  explicit *see* explicit memory(ies)
  hippocampus in, 130
  implicit *see* implicit memory(ies)
  long-term, 17, 18*f*
  nondeclarative, 17, 20
  procedural *see* procedural memory
  short-term *vs.* working, 10
  sleep effects on, 134
  state-based, 21, 230
  threat-based, 200
  working, xxi, 10, 142
memory evolution
  from memory reconsolidation to, 133–34
memory integration, 199–202
  steps in, 200–2
memory reconsolidation
  to memory evolution, 133–34
memory systems. *see also specific types*
  hippocampal-frontal, 200
  long-term *see* long-term memory systems
  mental operating networks and, 17
  therapeutic reconsolidation of, 199
mental health
  ACE study ramifications on, xxii–xxiii
  beliefs and, 248
  chronic inflammation effects on, xxiv
  computer games effects on, xxvi
  constructing solution-oriented stories in, 15
  diet impact on, xxv
  excessive fat cells effects on, 100–2
  exercise in improving, 126
  forgiveness, compassion, and gratitude impact on, xxx
  gambling effects on, xxvi
  gratitude effects on, xxx, 252–53
  in health behaviors, 93
  integrated approach to, 64
  integrated psychotherapy impact on, xxvi
  nurturing impact on, xxii
  placebo effects and, 245–46
  psychoneuroimmunology in identifying dysregulations of, 93
  self-care practices in undermining, xxiv–xxv
  telomere length impact on, 115
mental operating networks, xxi–xxii, 5–15. *see also specific types, e.g.,* salience network
  activity related to, 260
  attachment and, 49–51
  balancing of, 15–17
  case examples, 6–7, 9, 11–12, 14–15
  central executive network, 6
  components of, 2
  contemplative attention and, 258–60, 259*f*
  in decision making, 21
  DMN, 6, 12–15, 13*f*
  executive network *see* executive network
  imbalance of, 60–61

long-term memory and, 18*f*
meditation and, 258–60, 259*f*
memory systems and, 17
rebalancing of, 237–42
salience network *see* salience network
in "self"-organizing, 267
in sense of self, 2
sleep deprivation effects on, 133
switching between, 15–16, 50–51
therapy related to, 16
working together, 15–16
Merton, T., 256
"messiness"
reparation for, 41
meta-awareness, 257
"metabolic/mitochondrial enzyme genes," 128
"metabolic-priority genes," 128
metabolic stress, 128
metabolic syndrome
case example, 92
characteristics of, 102
hypocorticolism and, 173–74
prevalence of, xxiv, 102
metabolic toxemia, 110
metacognitive models
from cognitive behavioral therapy to, 185
methylation, 73
DNA, 79
GDNF gene effects on, 85
gene expression through, xxiii
N-methyl-D-aspartate (NDMA) expression, 43
Met/Met genotype, 221
metta meditation, 262
microbe(s)
diversity of, 109
function of, 109
in norepinephrine modification, 113
microbiome(s), 108
microbiota
gut, 108–9
health and, 108
microglia, 98
PICs released from, 98
microorganisms
in gut, 107–9
mid-insula, 7, 9
mid-sleep cycle awakening, 136–37
mild detachment, 209–10
mind
beliefs effects on, 244–53
composition of, 5–15 *see also* mental operating networks
gaps between brain and, xix
as meaning-making process, xx–xxi
operating networks of *see* mental operative networks
placebos effects on, 244–53
"self"-organizing, 267–69
mind–body, xxiii–xxiv
health and, 91–114 see also health
placebo response as shift in, 246–47
mind–body chaos, 164–75
mind–body class
at Kaiser Permanente Medical Centers, 264–65
mind–body medicine
in medical centers, 264–65
mind–brain–gene feedback loops, xviii, 68, 269
epigenetics in, 58*f*
from genetic reductionism to, 89
habit and motivation in, 138*f*
psychoneuroimmunology in, 90*f*
self-organization in, xxxii*f*
self-regulation in, 114*f*
social self in, 26*f*
stress effects on, 105
mind–brain–gene interactions, 111–13
determinants of, 67
stress effects on, 112
top-down and bottom-up pathways in, 112
mindfulness
for depression, 241
mindfulness-based cognitive therapy, 265–66
mindfulness-based stress reduction (MBSR), xxx, 263–64
for chronic pain, 266
effectiveness of, 264–65
immune system effects of, 265
indications for, 264, 266
in moderating stress response systems, 264
for obsessive-compulsive disorder, 266

mindfulness practices. *see also specific types*
health effects of, xxx
mind in time, xxix–xxx, 243–69
"mind reading"
oxytocin in, 43–44
mind's operating networks. *see* mental operating networks
mirror neurons, 31
moderate detachment, 209–10
moderate matching
perfect matching *vs.*, 38
moment(s)
being present in, xxi
present, 12
monastic orders, 256
mood
BDNF in, 100
chronic inflammation effects on, xxiv
sleep impact on, xxv
mood disorders
insecure attachment and, 48
mother(s)
infants of depressed, 37
mother–infant dyad, 41
motivation, 139–61
addiction and, 154–59, 158f
amygdala in, 147, 148
building on, 161
to change, 160
dopamine in, 142
extrinsic, 161
habits effect on, xxv–xxvi
implicit memory system and, 141
intrinsic, 161
kindling through bottom-up and top-down approaches, 159–61
in mind–brain–gene feedback loops, 138f
nucleus accumbens in, 147, 148
optimizing, 152
reward-seeking system and, 140–41
searching for, 140–53, 144f
motivational interviewing, 160–61
active listening in, 57
elements in, 161
motivational system
amygdala in, 147, 148
nucleus accumbens in, 147, 148
structures in, 235
motivation circuit, 236
motivation–pleasure circuit, 141–46, 144*f*
case examples, 141–46
movement
in self-maintenance, 125–31 *see also* exercise(s)
striatum-driven, 236
mucus layer
health of, 111
mucus production
regulation of, 111
muscle tension
stress and, 129
mystic(s)
Medieval Christian, 256

narcissistic children
causes of, 39
Narcotics Anonymous, 148, 151, 257
narration(s)
therapeutic, 53–55
narrative(s)
co-constructive, 54–55, 202
"somatic," 212
natural killer cells, 168
natural selection
Darwin's theory of, 68–69
Navajo Reservation
relationship building on, 57
negative feedback loops
function of, xx
negativity
"uprooting," 261
neglect
emotional, 42, 60–61
nerve growth factor (NGF), 85
depression and, 220
"nervous gut feelings," 107
"nervous stomach," 111
neural circuit, 236
neural plasticity
BDNF in, 100
neurobiology
interpersonal, xix
neurodegenerative diseases
chronic inflammation and, 97
PICs and, 110
types of, 97
neuroendocrine system, 79–80
dysregulation of, 166

neurofeedback instruments
  in attaining deeper states of relaxation, 264
neurogenesis
  BDNF in, 100
  blocking of, 110–11
neurohormone(s)
  epigenetic changes to, 86
  exercise effects on, 129
neuron(s)
  dopamine, 142–46, 144*f*, 152
  medium spiny, 147–52
  mirror, 31
neuropeptide(s)
  exercise effects on, 129
neuroplasticity
  for long-term memory systems, 17
neuroscience
  psychotherapy with, xix
neurotransmitter(s), 108, 123
  altering levels of, xxiv
  amino acids as building blocks for, 123
  exercise effects on, 129
  inhibiting, 76
neurotrophic factors
  in brain enhancement and repair, 130
  depression related to, 220–22
  described, 220
  glial cell–derived, 130
  types of, 220
neutrophil(s), 96
NGF. *see* nerve growth factor (NGF)
NGF gene
  alteration of, 85–86
Niebuhr, R., 254
NMDA expression. *see* *N*-methyl-D-aspartate (NMDA) expression
nocebo effects, 244–53
  defined, 246
  described, 246–47
nocebo–placebo, 246–48
nocebo response
  described, 246–47
nondeclarative memory, 17, 21
  recalled through salience network, 20
nondrug addictions, 155
non-sleep-time daydreaming
  time spent in, 12
norepinephrine, 123, 131, 167, 168, 171, 172
  bacterial pathogens growth related to, 113
  in fight-or-flight response, 168
  gut microbes in modification of, 113
  release of, 165
  trauma and, 134
not engaging
  time spent, 12
nourishment
  emotional, 39
nucleus accumbens, 125, 141, 143, 235, 236
  function of, 9
  in motivational system, 147, 148
nurturance, 3
  lack of emotional, 36
  mental health effects of, xxii
  stress tolerance effects of, 59
nurtured nature
  social self and, 28–29, 29*f*

obese people
  chronic inflammation in, 102
  IL-6 in, 102
  PICs in, 102
obesity. *see also* overweight
  Alzheimer's disease and, 101
  BDNF effects of, 101
  BMI in, 100–1
  CDC on, 100–1
  as chronic subclinical inflammatory condition, 100–2
  defined, 100
  dementia related to, 101
  described, 111
  impact on dysregulation of energy intake and expenditure, 101
  increase in, 126
  inflammation related to, 100
  leptin resistance related to, 100
  prevalence of, xxiv, 100
  telomeres shortened by, 100
  type 2 diabetes related to, 101
obsessive-compulsive disorder
  mindfulness and cognitive behavioral therapy for, 266
obstructive sleep apnea, 135
  CPAP device for, 135
offspring of mothers who suffered PTSD
  Holocaust-related, 84

O'Keefe, G., 261
Okinawan diet, 124, 125
oligofructose, 111
omega-3 essential fatty acids
  BDNF and, 122
  benefits of, 122
  brain functions and, 121–22
  excessive doses of, 122
omega-3 fish oil supplementation
  effects of, 121
open systems
  closed systems *vs.*, xxix
operating networks of mind. *see* mental operating networks
opioid(s)
  in attachment, 43
  endogenous, 43
opioid circuitry
  in enjoying, 142
opportunity(ies)
  managing, 204–9, 205f
optimism
  neuroscience of, 255
orientation response
  shift in attention involving, 214
overeating
  case example, 145–46
overweight. *see also* obesity
  case examples, 91–94, 97, 98, 100–4, 111–12
  prevalence of, xxiv, 100, 126
  sleep-disordered breathing and, 135
  WHO on, 100
oxidative stress
  BDNF effects of, 122
oxytocin
  in attachment, 43
  calming effect related to, 43–44
  epigenetic changes to, 86
  function of, 86
  in "mind reading," 43–44
  receptors for, 165–66
  as regulator of emotional and prosocial behaviors, 43
  in stress reduction, 43
oxytocin receptors
  in activating parasympathetic nervous system, 43
pain
  ACC and, 32
  anterior insula and, 32
  chronic, 266
  exercise in modulating, 166
  physical, 32, 231–32
  reduction of, 245
  social, 32, 231–32
Paleolithic diet, 117
panic
  formation of, xxvii
panic attack(s), 186–88
panic disorder, 186
  body sensations accompanying, 5
panic/grief, 3
Panksepp, J., 3
parasympathetic nervous system, 165
  immune system relationship with, 165
  oxytocin receptors in activating, 43
  vagus nerve system of, 33, 34f
paraventricular nucleus, 198
parent(s)
  "good-enough," 38–41
parental depression
  impact on infants, 35–38
parent–child dyad
  as self-organizing system, 41
  stories in, 53
parent–infant dyad
  quality of reciprocity within, 39
parenting
  "good-enough," 38–41
passive-aggressive behaviors, 39
pathophysiological syndromes
  early stress and deprivation resulting in, 112–13
Pavlov, 212–13
perceived stress
  telomere length and, 87
perfect matching
  moderate matching vs., 38
personal relevance, 147
pessimist(s)
  causes of, 39
Peyer's patches, 106–7
PFC. *see* prefrontal cortex (PFC)
phenylalanine, 123
phobia(s), 186
  formation of, xxvii

physical activity
  prevalence of, 126
  reduced, 128
physical health
  in health behaviors, 93
  psychoneuroimmunology in identifying dysregulations of, 93
physical illnesses
  prevalence of, 93
physical inactivity
  chronic illnesses related to, 126
physical pain
  ACC in detecting, 231–32
  social pain overlapping with, 32
PICs. *see* pro-inflammatory cytokines (PICs)
pineal gland, 135
placebo(s)
  beliefs and, 244–53
  defined, 244
  described, 244
  negative effects from, 246
  in pain reduction, 245
placebo effect(s), 244–53
  in brain, xxix
  case example, 243–53
  described, 248
  factors increasing, 245
  illustrating how beliefs can alleviate suffering and change biology, 244
  interactions with social and biological factors, 248
  as "in the head," 246
  shift in mental health treatment related to, 245–46
  theoretical support for, 244
  treatments with benefits of, 244
placebo–nocebo
  beliefs and, 246–48
placebo response
  case example, 246
  incidence of, 245
  as "meaning response," 247
  as shift in mind–body, 246–47
"plants plus" diet, 124
play, 3
  face-to-face, 56–57
  insulin-like growth-factor and, 85
playfulness
  therapy enhanced by, 85
"playing possum," 209
pleasure
  subjective experience of, 9
polyvagal system, 33
poor health
  chronic stress and, 174–75
  synergistic depressive effects of, 229f
  undetermined stress tolerance related to, 174–75
poor self-care
  synergistic depressive effects of, 229*f*
positive attitudes, 250–51
positive behaviors, 250–51
positive emotional states
  gratitude and, 252
positive feedback loops, xx
positive psychology
  attitudes and, 249–50
  theoretical roots of, 250–53
positive relationships
  health related to, xxii
posterior parietal cortex (PPC), 11, 11*f*
postpartum depression
  case example, 35–36
  described, 36–37
  impact on infants, 35–38
posttraumatic growth
  therapy and, 214–16
posttraumatic stress disorder (PTSD)
  acute stress disorders becoming, 134
  case example, 191–216
  causes of, xxviii
  disorders related to, 65
  DMN and right amygdala in symptoms of, 16
  Holocaust-related, 84
  long-term dysregulations associated with, 198–99
  neuroimaging findings related to, 196
  prevalence of, 195–202
  prevention of, 195–202
  risk factors for, 192
  stabilization of, 195–202
Power of Persuasion, 247
PPC. *see* posterior parietal cortex (PPC)
prayer(s)
  calming factors in, 257–58
  contemplative, 256

prefrontal cortex (PFC), xxvii
in changing bad habits into good ones, 156–57
in comparing knowledge, 22
conscious reasoning in, 145
development of, 147
dorsolateral, 10–11, 11f
executive functions of, 10–12, 11f
in forming and modifying habits, 158–59, 158f
schizophrenia impact on, 77
shoring up after trauma, 197–98
size of, 9
prefrontal deficits
impulse control problems and affect regulatory problems associated with, 147
pregnancy
stress hormones during, 84
pregnant women
at time of World Trade Center attacks, 84
preoccupied pattern
case example, 48
present moment
time spent in, 12
priming
for autostress disorders, 176–77
procedural memory, xxvi
described, 18–19
habits and, 19
providing clues for explicit memory, 20
as subsystem of implicit memory system, 18–20
procedural memory system
implicit-nonconscious, 211
programmed cell death, 172
pro-inflammatory cytokines (PICs), 96–97, 117
abnormal levels of, 223, 224
astrocytes releasing, 98
basal ganglia effects of, 225–26
depression and, 226–27
DMN and salience network effects of, 227
in gut, 107, 110
hippocampus effects of, 226–27
hypothalamus effects of, 226–27
illness effects on, 105
inflammation related to, xxiv
microglia releasing, 98
neurodegenerative disorders related to, 110
in obese people, 102
overactivation of, 121
overexpressed, 99
stress in activation of, 103
promoter(s), 67
prosocial behaviors
oxytocin as regulator of, 43
protein(s)
COMT, 77
described, 72
protein-coding genes
numbers of, 67
protein transcription factors, 72
Proteobacteria, 110
psychology
positive, 249–53
"pseudo-depression," 10
psychoimmunology
focus on interactions among mind, brain, and immune system in, xix
psychological adversity
physical health problems related to, 61
psychological disorders
autoimmune disorders as bidirectional causal interactions with, 94, 102–3
inflammation and, 97
obesity and, 100
telomere length and, 88
vulnerability to, 73–74
psychological experiences
gene expression effects of, 61
psychological factors
bidirectional causal relationships with, 65
psychological feedback loops
environmental impact on, 71
psychological rigidity
depression as being stuck in, xxix
psychological stress
in children, 64
psychological stressors
bad health related to, 163–64
psychoneuroimmunology
components of, 93
in identifying interrelated mental and physical dysregulations, 93

in mind–brain–gene feedback loops, 90f
of social support, 34–35
psychosis(es)
COMT gene and, 77
psychotherapy
as behavioral health, 92
beliefs about, 247–48
for depression, 240–41
described, x
directional changes in, x–xi
epigenetics in revolutionizing, 70, 71
gene expression effects of, xix, 59
health care interdependence with, xviii–xix
in influencing gene–environment interactions, 59
integrated, xxvi
mental health effects of, xxvi
neurophysiological changes resulting from, 207
neuroscience with, xix
psychotherapy training
cultural competence in, 57
PTSD. *see* posttraumatic stress disorder (PTSD)

quality of care early in life
epigenetics on effects of, 59
quigong
effectiveness of, 265

rage, 3
rapid eye movement (REM) sleep, 134
reappraisal
cognitive, 160
reasoning
conscious, 145
in PFC, 145
reciprocal self-awareness, 51–53
reciprocity
in infant–parent dyad, 39
reconsolidation
memory, 133–34
reductionism
genetic *see* genetic reductionism
reflection(s)
self-referential, 13
reflex(es)
"inflammatory," 165
stress-induced autonomic-inflammatory, 198
refugee(s)
prevalence of, 191
relatedness
interpersonal, 52
relationship(s)
close, xxii
internal working model of, 41
positive, xxii
shared implicit, 41
transference, 52
relationship building
on Navajo Reservation, 57
relaxation exercises
calming factors in, 257–58
for GAD, 182–83
relevance
amygdala as detector of, 22
assessing of, 9
personal, 147
REM sleep. *see* rapid eye movement (REM) sleep
reparation(s)
high-sensitivity, 40
low-sensitivity, 40
for "messiness," 41
resilience
enhanced nurturing promoting, 86–87
factors promoting, 84–87
neurochemistry of, 84–87
stress, 168–69
resiliency
described, 170
as durability, 170
"resting state" activity
abnormal, 133
reward(s)
alcohol as, 154
reward circuit(s), 140–41
in addiction-related work, 160
in dopamine release, 146
effort-driven, 236
salience network and, 141
reward-seeking system
in alcoholism, 154
motivation and, 140–41

reward system
  appetite cycle and, 125
  refined subcortical, 148
rhythm(s)
  ultradian, 12
right hemisphere
  during development, 28–29, 29*f*
right/left balancing
  bottom-up approaches and, 197–98
rigidity
  psychological, xxix
  transcending, xxviii–xxix, 217–42 *see also* transcending rigidity
Risky Situation (RS), 40–41
"rock"
  cocaine in form of, 154
Rohr, R., Friar, 261
Rossi, E., xix, 78
Rough-and-Tumble Play (RTP) paradigm, 41
RS. *see* Risky Situation (RS)
RTP paradigm. *see* Rough-and-Tumble Play (RTP) paradigm
rumination, xxi
  time spent in, 12
runner's high, 129, 166
"rush" of excitement, 143–45

sadness, 3
  as CER, 208
Saint Augustine, 256
salience network, xxi, 6–9, 8*f*
  components of, 7–8, 8*f*
  DMN with, 50, 238–39
  executive network with, 146–47, 188–89
  function of, 15–16
  implicit memories recalled through, 20
  long-term memory systems in, 21
  in meditation, 258–60, 259*f*
  PICs effects on, 227
  in promoting gut or intuitive feelings, 32
  reward circuit and, 141
  sleep deprivation effects on, 133
saturated fat
  diet high in, 120
"scheduling worrying time," 184–85
schizophrenia
  "latent," 77
  PFC effects of, 77
secondary emotions
  tormenting nature of, 203
  trauma-related, 203–4
secondary reinforcers
  in extrinsic motivation, 161
"second brain"
  gut as, 105–6
second-messenger system
  essential fatty acids in helping brain facilitate, 123
secure attachment
  as buffer to effects of trauma, 81f, 83–84
  in childhood, 47
  cycle of epigenetic development of stress tolerance by, 81f
  sensitivity in, 39
SEEDS (social, exercise, education, diet, and sleep factors)
  in self-maintenance, 137
seeking (expectancy), 3
self
  components of subjective organization of, 2
  feedback loops as self-organizing systems within, xx
  feelings in awareness of, 267
  memory systems in background fabric of, 17
  sense of *see* sense of self
  "sentient," 6
  social *see* social self
  somatic sense of, 2
self-agency, 15
self-awareness
  reciprocal, 51–53
self-care
  ACEs impact on, 63–64
  poor, 229*f*
self-care practices
  in undermining health and mental health, xxiv–xxv
self-compassion, 254–55
  benefits of, 255
  in contemplative meditation, 262
self-criticism, 254
self-kindness, 254
self-maintenance, xxiv–xxv, 115–38
  case example, 115–20

diet in, 115–25
movement in, 125–31
SEEDS in, 137
self-maintenance system(s), 115
"self"-organization, xxi–xxii, 1–26
balancing network in, 15–17
benefits of, 267–69
body in, 3–4
described, xx
emotional neglect shaping pattern of, 42
emotions in, 3–5
factors contributing to, 38
implicit memories forming fabric of, 20
long-term memory systems in, 17–22, 18*f*
mental operating networks in, 5–15 *see also specific types and* mental operating networks
in mind–brain–gene feedback loops, xxxii*f*
posttraumatic, 215
progression of, 39
thoughts and, 3–4
top-down and bottom-up interactions in, 4–5
"self"-organizing mind, 267–69
mental operating networks in, 267
"self"-organizing networks, 2
"self"-organizing process, xx
mind as, 268–69
self-organizing system
parent–child dyad as, 41
self-preservation, 195
self-referential information systems, xxi–xxii
defined, 13
self-referential reflections, 13
self-referential thought, xxi
self-regulation
in mind–brain–gene feedback loops, 114*f*
social self and, 41–44
self-soothing
attachment and, 79
Selye, H., 163, 169
semantic memory
described, 20–21
explicit, 20–21
sensation(s)
body, 211–14
bottom-up, xxix
as side effects, 248
somatic, xxi, 7–9, 8*f*
sense of self
mental networks in, 2
somatic, 2
underdeveloped, 27–33
sense of spirituality, 251–52
sensitivity
anxiety, 129, 186–87
in secure attachment, 39
sensitization
for autostress disorders, 176–77
Sensorimotor Integration, xxviii, 212
"sentient self," 6
separation
social self and, 44–47
Serenity Prayer, 254
serotonin, 123, 124
good nurturing stimulates release of, 78
serotonin levels
exercise effects on, 129
serotonin-transporter gene (5-HTTLRR), 75
short version of, 74–75, 219
in vulnerability to depression or GAD, 75–76
sexual excitement, 3
sexual repression theory, 61
shared implicit relationship, 41
Shenpen, L., 261–62
shift in attention
orientation response in, 214
Shore, A., xix
short-chain fatty acids, 125
short-term inflammation
chronic inflammation *vs.*, 96
short-term memory
working memory *vs.*, 10
short-term stress, 103
short version of serotonin transporter gene
ACEs and, 74–75
Shto takoe?, 213
shutting down, 193
sickness behavior, 224–27
in adaptation, 225
characteristics of, 224–27
chronic inflammation and, 224–27
symptoms of, 224–26

side effects
  sensations as, 248
side-to-side approaches
  in autostress disorders, 180–81
side-to-side frame of reference, 177–78
Siegel, D.J., xix
"silencing," 72–73
simple carbohydrates
  BDNF effects of, 119
sinus infections
  case example, 105, 107
sleep, 131–35
  alcohol in suppression of, 136–37
  in brain maintenance, 132
  cognitive effects of, xxv
  "core," 132
  deprivation of, 132–33
  factors associated with impaired, xxv
  fragmented, 134
  functions of, 132
  immune system effects of, 105
  memory effects of, 134
  mood effects of, xxv
  need for, 131–35
  optimal amount of, 132
  percentage of life in, 131
  quality of, 131–35
  REM, 134
  slow-wave, 134, 136–37
  toxins released during, 131–32
  trauma effects on, 134
sleep apnea
  CPAP device for, 135
  obstructive, 135
sleep deprivation, 134
  adverse effects of, 131–35
  appetite effects of, 133
  disorders and conditions related to, 132–33
  DMN connectivity effects of, 133
  emotional memories and, 134
  immune system effects of, 134
  memory effects of, 134
  mental operating networks effects of, 133
sleep-disordered breathing, 135
sleep disturbances, 134
  CRP effects of, 135
  inflammation and, 135
sleep hygiene techniques, 131
"sleep maintenance" insomnia, 136–37
slow track
  to amygdala, 22–25, 23f
  speeding up, 188–89
slow-wave sleep, 134
  alcohol in suppression of, 136–37
social brain
  components of, 30–31
social brain circuits, 30–31
social brain networks, 30–31
  kindling of, xxii
social competence
  requirements for, 33
social engagement system, 33, 193–94
social factors
  gene expression and, 78, 82
  placebo effects interacting with, 248
social genomics
  field of, 78
social joy, 3
socially avoidant coping methods
  case example, 42
social pain
  ACC in detecting, 231–32
  physical pain overlapping with, 32
social self, xxii, 27–58
  attachment and, 47–51
  attachment styles and, 41–44
  coevolving and, 55
  cultivating empathy and, 31–33
  cultural variations and, 55–57
  "good-enough" parenting and, 38–41
  maternal depression effects on, 35–38
  mental operating networks and, 49–51
  in mind–brain–gene feedback loops, 26*f*
  nurtured nature and, 28–29, 29*f*
  reciprocal self-awareness and, 51–53
  self-regulation and, 41–44
  separation, loss, and building trust related to, 44–47
  social brain networks and, 30–31
  still face paradigm impact on, 35–36
  storytelling and, 53–55
  therapeutic narrations and, 53–55
  vagal brake and, 33–35, 34*f*
social stress
  depression and, 230–32

social stressor(s), 193
social support
  psychoneuroimmunology of, 34–35
  stress reduction related to, 43
social withdrawal
  chronic inflammation effects on, xxiv
  synergistic depressive effects of, 229f
soft drinks
  in Western diet, 119
Solomon, G.F., xix
solution-oriented stories
  in mental health, 15
somatic exercises
  for GAD, 182–83
Somatic Experiencing, xxviii, 212
somatic grounding, 238
somatic medicine, 61
"somatic narrative," 212
somatic sensations, xxi, 7–9, 8*f*
somatic sense of self
  evolving experiences of individuality emerge from, 2
somatic stimulation, 214
somatic therapies, 211–14
spindle cells, 9, 30–31
spiny neurons
  greater gain related to, 147–52
spirituality
  sense of, 251–52
spoiled children
  causes of, 39
stability
  affect, xxii
state-based memory, 21, 230
still face paradigm, 35–36
stimulation
  somatic, 214
stomach
  "nervous," 111
story(ies)
  benefits of, 53
  in interaction between parents and child, 53
  solution-centered, 15
"story brain." *see* default-mode network (DMN)
storytelling, 53–55
  benefits of, 53
  functions of, 53
strength(s)
  building of, 156–57
stress, 163–89. *see also* autostress
  acute, 172–75 *see also* acute stress
  acute and chronic *vs.* mild to moderate, 170
  anxiety related to, xxvii
  attachment styles in coping with, 42
  auto-, 163–89 *see also* autostress disorders
  autostress and, 163–89
  bodily effects of, 198
  case example, 163–89
  chronic *see* chronic stress
  CRH levels and, 228
  depression and, 219, 228–29
  early, 112–13
  effects on mind–brain–gene feedback loops, 105
  IL-6 and, 168
  immune system disruptions and, 166–68, 167*f*
  immune system effects of, 103
  impact on extracellular and intracellular signaling and gene-expression pathways in brain, 76
  impact on mind–brain–gene interactions, 112
  inflammation and, 103
  interpersonal, 42
  limitations of term, 169
  metabolic, 128
  muscle tension related to, 129
  oxidative, 122
  perceived, 87
  PICs activating, 103
  psychological, 64
  severe, 228
  short-term, 103
  social, 230–32
  synergistic depressive effects of, 229*f*
  telomere length and, xxiii
  transgenerational effects of, 84
  transitioning into autostress disorders, 177
  traumatic, 173
stress–depression synergy, 227–30, 229*f*
stressful events
  gene functioning effects of, 76

stress hormones
high levels during pregnancy, 84
stress-induced autonomic-inflammatory reflex, 198
stress–inflammation connection, 103–5
stressor(s)
acute, 168
bad health related to, 163–64
psychological, 163–64
social, 193
stress-provoking stimuli
structures in exposure to, 179–80
stress reduction
mindfulness-based *see* mindfulness-based stress reduction (MBSR)
oxytocin in, 43
social support in, 43
stress-related learning
emotion/cognition interface in, 76
stress relievers
alcohol as, 154
types of, 139
stress resilience, 168–69
stress response(s)
case example, 163–89
immune responses modified by, 164
to threat, 167f
stress-response genes, 128
stress response subsystems, 79–80
stress response systems
autostress disorders impact on, xxvii
breakdown of, 193–94
damping down, 180–81
MBSR in moderating, 264
stress system(s)
feedback loops between homeostatic system and, 164
stress systems
multidimensional, xxvi–xvii
stress tolerance
cycle of epigenetic development of, 81f
diet impact on, 116, 124
GDNF gene and, 85
genes impact on, xxiii
good early nurturance effects on, 59
underdevelopment in, 79–80
undetermined, 174–75
striatum, 235
striatum-driven movement, 236
"stumbling on happiness," 250
subcortical reward system
refined, 148
subjectivity
emergence of, xxi–xxii
substance abuse
epigenetics and, 76–78
suicide
ACEs and, 79
ACE study score and, 79
CDC on, 224
depression and, 224
low opioid receptor activity in brain and, 79
Sylvian fissure, 7
synchronization
attention, 264
synchronized activity, 264

tai chi
effectiveness of, 265
talk
change, 160–61
counterchange, 160
tapping therapies, 212
telomerase, 87, 88
described, xxiii
telomere(s), 87–88
described, xxiii, 87
Mediterranean diet in maintenance of, 124
obesity effects on, 100
shortening of, xxiv, 87, 115
sleep effects of, 132
telomere length
as biomarker for health and longevity, 87
of caregivers, 87
chronic stress and, 87–88
depression related to, xxiii, 87–88
factors associated with, 87–88
lifestyle behaviors related to, xxiii
mental health effects of, 115
perceived stress and, 87
psychotherapy in promoting longer, 88
stress related to, xxiii
temperature
regulation of, 136–37

tension
  muscle, 129
The Descent of Man, 27
"the eye of the heart," 256
"the material me." *see* salience network
theory(ies). *see specific types*
The Phenomenon of Man, 269
therapeutic narrations, 53–55
therapeutic window
  neurobiological mechanisms of, 206–7
therapy. *see also specific types*
  body sensations in, 211–14
  exercise and, 130–31
  implicit feeling of, 53
  mental operating networks–related, 16
  posttraumatic growth as goal of, 214–16
The Selfish Gene, 68
"the survival of the fittest," 68
thought(s)
  automatic, 189
  emotions and, 3–4
  intrusive trauma-related, 199
  self-referential, xxi
  worrisome, 185
thought diffusion, 185
threat(s)
  amygdala in detection of, 176
  feedback loops underlying stress and immune responses to, 167f
threat-based memory, 200
threat detection system, 188–89
Tibetan meditation, 258–59
time
  feelings in awareness of, 267
  mind in, xxix–xxx, 243–69
TNF. *see* tumor necrosis factor (TNF)
"too much stress on this bridge," 169
top-down control–bottom-up changes, 147–48
top-down control–bottom-up circuits
  maintaining and enhancing, 10
top-down patterns, xxix
  in autostress disorders, 180–81
  emotions constructed from, 7–8, 8f
  emotions-related, 4–5
  kindling motivation through, 159–61
  in mind–brain–gene interactions, 112
  unbalances in, 176
top-down willpower, 156
toxemia
  metabolic, 110
toxin(s)
  released during sleep, 131–32
transcending rigidity, 217–42
  case example, 217–42
  depression and, 218–37 *see also* depression
transcription factors
  described, 72
  protein, 72
transduction, 72
trans-fatty acids
  destructive effects of, 120–21
transference relationship, 52
transgenerational trauma
  behavior–gene interactions and, 82–84
trauma
  complex, 198
  depression related to, 223
  developing meaning after, 214–16
  early-life, 223
  feedforward loop related to symptoms of, 196–97
  fragmented sleep resulting from, 134
  growth after, 214–16
  impact on sleep, 134
  long-term dysregulations associated with, 198–99
  norepinephrine and, 134
  secure attachment as buffer to effects of, 81*f*, 83–84
  "self"-organization after, 215
  serious illnesses and death among survivors of, 65
  shoring up PFC after, 197–98
  therapeutic phases applied to, 197
  transgenerational, 82–84
trauma-induced responses, xxviii
trauma-related thoughts and feelings
  intrusive, 199
trauma spectrum, xxviii, 191–216
  avoidance, detachment, and dissociation, 209–14
  breakdown of systems in, 193–95
  case examples, 191–216
  managing window of tolerance and opportunity, 204–9, 205f

trauma spectrum (*continued*)
secondary emotions in, 203–4
stabilizing and preventing PTSD, 195–202
traumatic event(s)
bodily effects of, 195
defensive behaviors related to, 193–94
traumatic stress
hypoactivation of HPA axis related to, 173
triglycerides, 121
triune brain theory, 3
Trump, D., Pres., 250
trust
building of, 44–47
social self and, 44–47
L-tryptophan, 123, 124
tumor necrosis factor (TNF)
abnormal levels of, 223
tumor necrosis factor (TNF) alpha, 96
Twain, M., 234
twelve-step programs, 148, 151. *see also specific programs*
twin(s)
behavior–gene interactions in, 59–90
type 2 diabetes
case example, 92, 104
cortisol levels and, 174–75
depression related to, 102, 224
IL-6 and, 102
increase in, 126
memory effects of, 119
obesity and, 101
prevalence of, xxiv, 101
Tzu, C., 250

ultradian rhythms, 12
underdeveloped sense of self
case example, 27–33
underdevelopment
in stress tolerance, 79–80
"uprooting negativity"
methods for, 261
uricase, 117
US National Institute of Health, 108

vagal brake, 33–35, 34*f*
vagus nerve system, 165
of parasympathetic nervous system, 33, 34f
Val/Val genotype, 221
Vipassana meditation, 259
vulnerability
to ACEs, 74
causation *vs.*, 74
to depression, 75–76
to GAD, 75–76
to psychological disorders, 73–74
"vulnerability gene," 75

"walking on eggshells," 192
wanting
dopamine circuits and working memory related to, 142
Western diet
composition of, 116–25
Whitehall Study, 132
WHO. *see* World Health Organization (WHO)
willpower
inadequacy of, 156–57
top-down, 156
window of tolerance
managing, 204–9, 205f
Winnocott, D., 38, 39
withdrawal behaviors
depression and, 234–37
Woman's Worth, 261
working memory, xxi
short-term memory *vs.*, 10
in wanting, 142
World Health Organization (WHO)
on cancer related to diet high in saturated fat, 120
on overweight, 100
on usefulness of ACE questions in health survey, 64
World Trade Center attacks
mothers pregnant at time of, 84
worrisome thoughts, 185
worry
as cognitive avoidance, 183–85
"worrying time"
scheduling, 184–85

yoga
effectiveness of, 265

Zen meditation, 259

Also available from

THE NORTON SERIES ON INTERPERSONAL NEUROBIOLOGY

The Birth of Intersubjectivity: Psychodynamics, Neurobiology, and the Self
Massimo Ammaniti, Vittorio Gallese

Neurobiology for Clinical Social Work: Theory and Practice (Second Edition)
Jeffrey S. Applegate, Janet R. Shapiro

Mind–Brain–Gene
John B. Arden

The Heart of Trauma: Healing the Embodied Brain in the Context of Relationships
Bonnie Badenoch

Being a Brain-Wise Therapist: A Practical Guide to Interpersonal Neurobiology
Bonnie Badenoch

The Brain-Savvy Therapist's Workbook
Bonnie Badenoch

The Neurobiology of Attachment-Focused Therapy
Jonathan Baylin, Daniel A. Hughes

Coping with Trauma-Related Dissociation: Skills Training for Patients and Therapists
Suzette Boon, Kathy Steele, and Onno van der Hart

Neurobiologically Informed Trauma Therapy with Children and Adolescents: Understanding Mechanisms of Change
Linda Chapman

Intensive Psychotherapy for Persistent Dissociative Processes: The Fear of Feeling Real
Richard A. Chefetz

Timeless: Nature's Formula for Health and Longevity
Louis Cozolino

The Neuroscience of Human Relationships: Attachment and the Developing Social Brain (Second Edition)
Louis Cozolino

The Neuroscience of Psychotherapy: Healing the Social Brain (Second Edition)
Louis Cozolino

Why Therapy Works: Using Our Minds to Change Our Brains
Louis Cozolino

From Axons to Identity: Neurological Explorations of the Nature of the Self
Todd E. Feinberg

Loving with the Brain in Mind: Neurobiology and Couple Therapy
Mona DeKoven Fishbane

Body Sense: The Science and Practice of Embodied Self-Awareness
Alan Fogel

The Healing Power of Emotion: Affective Neuroscience, Development & Clinical Practice
Diana Fosha, Daniel J. Siegel, Marion Solomon

Healing the Traumatized Self: Consciousness, Neuroscience, Treatment
Paul Frewen, Ruth Lanius

The Neuropsychology of the Unconscious: Integrating Brain and Mind in Psychotherapy
Efrat Ginot

10 Principles for Doing Effective Couples Therapy
Julie Schwartz Gottman and John M. Gottman

The Impact of Attachment
Susan Hart

Art Therapy and the Neuroscience of Relationships, Creativity, and Resiliency: Skills and Practices
Noah Hass-Cohen and Joanna Clyde Findlay

Affect Regulation Theory: A Clinical Model
Daniel Hill

Brain-Based Parenting: The Neuroscience of Caregiving for Healthy Attachment
Daniel A. Hughes, Jonathan Baylin

Sex Addiction as Affect Dysregulation: A Neurobiologically Informed Holistic Treatment
Alexandra Katehakis

The Interpersonal Neurobiology of Play: Brain-Building Interventions for Emotional Well-Being
Theresa A. Kestly

Self-Agency in Psychotherapy: Attachment, Autonomy, and Intimacy
Jean Knox

Infant/Child Mental Health, Early Intervention, and Relationship-Based Therapies: A Neurorelational Framework for Interdisciplinary Practice
Connie Lillas, Janiece Turnbull

Play and Creativity in Psychotherapy
Terry Marks-Tarlow, Marion Solomon, Daniel J. Siegel

Clinical Intuition in Psychotherapy: The Neurobiology of Embodied Response
Terry Marks-Tarlow

Awakening Clinical Intuition: An Experiential Workbook for Psychotherapists
Terry Marks-Tarlow

A Dissociation Model of Borderline Personality Disorder
Russell Meares

Borderline Personality Disorder and the Conversational Model: A Clinician's Manual
Russell Meares

Neurobiology Essentials for Clinicians: What Every Therapist Needs to Know
Arlene Montgomery

Borderline Bodies: Affect Regulation Therapy for Personality Disorders
Clara Mucci

Neurobiology and the Development of Human Morality:
Evolution, Culture, and Wisdom
Darcia Narvaez

Brain Model & Puzzle: Anatomy & Functional Areas of the Brain
Norton Professional Books

Sensorimotor Psychotherapy: Interventions for Trauma and Attachment
Pat Ogden, Janina Fisher

Trauma and the Body: A Sensorimotor Approach to Psychotherapy
Pat Ogden, Kekuni Minton, Clare Pain

The Archaeology of Mind: Neuroevolutionary Origins of Human Emotions
Jaak Panksepp, Lucy Biven

The Polyvagal Theory: Neurophysiological Foundations of Emotions,
Attachment, Communication, and Self-regulation
Stephen W. Porges

The Pocket Guide to Polyvagal Theory: The Transformative Power of Feeling Safe
Stephen W. Porges

Foundational Concepts in Neuroscience: A Brain-Mind Odyssey
David E. Presti

Right Brain Psychotherapy
Allan N. Schore

The Development of the Unconscious Mind
Allan N. Schore

Affect Dysregulation and Disorders of the Self
Allan N. Schore

Affect Regulation and the Repair of the Self
Allan N. Schore

The Science of the Art of Psychotherapy
Allan N. Schore

Mind: A Journey to the Heart of Being Human
Daniel J. Siegel

The Mindful Brain: Reflection and Attunement in the Cultivation of Well-Being
Daniel J. Siegel

The Mindful Therapist: A Clinician's Guide to Mindsight and Neural Integration
Daniel J. Siegel

Pocket Guide to Interpersonal Neurobiology: An Integrative Handbook of the Mind
Daniel J. Siegel

Healing Moments in Psychotherapy
Daniel J. Siegel, Marion Solomon

Healing Trauma: Attachment, Mind, Body and Brain
Daniel J. Siegel, Marion Solomon

Love and War in Intimate Relationships: Connection, Disconnection, and Mutual Regulation in Couple Therapy
Marion Solomon, Stan Tatkin

How People Change: Relationships and Neuroplasticity in Psychotherapy
Marion Solomon and Daniel J. Siegel

The Present Moment in Psychotherapy and Everyday Life
Daniel N. Stern

The Neurobehavioral and Social-Emotional Development of Infants and Children
Ed Tronick

The Haunted Self: Structural Dissociation and the Treatment of Chronic Traumatization
Onno Van Der Hart, Ellert R. S. Nijenhuis, Kathy Steele

Prenatal Development and Parents' Lived Experiences: How Early Events Shape Our Psychophysiology and Relationships
Ann Diamond Weinstein

Changing Minds in Therapy: Emotion, Attachment, Trauma, and Neurobiology
Margaret Wilkinson

For all the latest books in the series, book details (including sample chapters), and to order online, please visit the Series webpage at wwnorton.com/Psych/IPNB Series